Ignition!

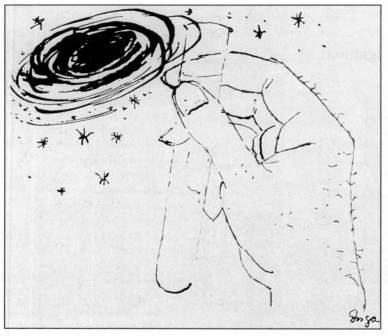

FIGURE 1 Engraving on plastic by Inga Pratt Clark, presented to Bob Border, Engineering Officer, NARTS, by the Propellant Division, 1959

FIGURE 2 This is what a test firing *should* look like. Note the mach diamonds in the exhaust stream. *U.S. Navy photo*

FIGURE 3 And this is what it may look like if something goes wrong. The same test cell, or its remains, is shown. *U.S. Navy photo*

Ignition!

An Informal History of Liquid Rocket Propellants

BY JOHN D. CLARK

> Those who cannot remember the past are condemned to repeat it.
>
> **GEORGE SANTAYANA**

Rutgers University Press

New Brunswick, Newark, and Camden, New Jersey, and London

Third paperback printing, 2018

Library of Congress Cataloging-in-Publication Data

Names: Clark, John D. (John Drury), 1907–1988, author.
Title: Ignition! : an informal history of liquid rocket propellants /
by John D. Clark.
Description: New Brunswick, New Jersey : Rutgers University Press, [2017] | Originally
published: New Brunswick, N.J. : Rutgers University Press, 1972. | Includes index.
Identifiers: LCCN 2017033845| ISBN 9780813507255 (cloth : alk. paper) |
ISBN 9780813595832 (pbk. : alk. paper)
Subjects: LCSH: Liquid propellants. | Liquid propellants—History. | Rockets
(Aeronautics)—Fuel—History.
Classification: LCC TL785 .C53 2017 | DDC 629.47/522—dc23
LC record available at https://lccn.loc.gov/2017033845

∞ The paper used in this publication meets the requirements of the American National
Standard for Information Sciences—Permanence of Paper for Printed Library Materials,
ANSI Z39.48–1992.

www.rutgersuniversitypress.org

Manufactured in the United States of America

This book is dedicated to my wife Inga, who heckled me into writing it with such wifely remarks as, "You talk a hell of a fine history. Now set yourself down in front of the typewriter—and write the damned thing!"

Contents

In Re John D. Clark

BY ISAAC ASIMOV

I first met John in 1942 when I came to Philadelphia to live. Oh, I had known *of* him before. Back in 1937, he had published a pair of science fiction shorts, "Minus Planet" and "Space Blister," which had hit me right between the eyes. The first one, in particular, was the earliest science fiction story I know of which dealt with "anti-matter" in realistic fashion.

Apparently, John was satisfied with that pair and didn't write any more s.f., kindly leaving room for lesser lights like myself.

In 1942, therefore, when I met him, I was ready to be awed. John, however, was not ready to awe. He was exactly what he has always been, completely friendly, completely self-unconscious, completely himself.

He was my friend when I needed friendship badly. America had just entered the war and I had come to Philadelphia to work for the Navy as a chemist. It was my first time away from home, ever, and I was barely twenty-two. I was utterly alone and his door was always open to me. I was frightened and he consoled me. I was sad and he cheered me.

For all his kindness, however, he could not always resist the impulse to take advantage of a greenhorn.

Every wall of his apartment was lined with books, floor to ceiling, and he loved displaying them to me. He explained that one wall was devoted to fiction, one to histories, one to books on military affairs and so on.

"Here," he said, "is the Bible." Then, with a solemn look on his face, he added, "I have it in the fiction section, you'll notice, under J."

"Why J?" I asked.

And John, delighted at the straight line, said, "J for Jehovah!"

But the years passed and our paths separated. The war ended and I returned to Columbia to go after my PhD (which John had already earned by the time I first met him) while he went into the happy business of designing rocket fuels.

Now it is clear that anyone working with rocket fuels is outstandingly mad. I don't mean garden-variety crazy or a merely raving lunatic. I mean a record-shattering exponent of far-out insanity.

There are, after all, some chemicals that explode shatteringly, some that flame ravenously, some that corrode hellishly, some that poison sneakily, and some that stink stenchily. As far as I know, though, only liquid rocket fuels have all these delightful properties combined into one delectable whole.

Well, John Clark worked with these miserable concoctions and survived all in one piece. What's more he ran a laboratory for seventeen years that played footsie with these liquids from Hell and never had a time-lost accident.

My own theory is that he made a deal with the Almighty. In return for Divine protection, John agreed to take the Bible out of the fiction section.

So read this book. You'll find out plenty about John and all the other sky-high crackpots who were in the field with him and you may even get (as I did) a glimpse of the heroic excitement that seemed to make it reasonable to cuddle with death every waking moment—to say nothing of learning a heck of a lot about the way in which the business of science is really conducted.

It is a story only John can tell so caustically well from the depths within.

Preface

Millions of words have been written about rocketry and space travel, and almost as many about the history and development of the rocket. But if anyone is curious about the parallel history and development of rocket propellants—the fuels and the oxidizers that make them go–he will find that there is no book which will tell him what he wants to know. There are a few texts which describe the propellants currently in use, but nowhere can he learn *why* these and not something else fuel Saturn V or Titan II, or SS-9. In this book I have tried to make that information available, and to tell the story of the development of liquid rocket propellants: the who, and when, and where and how and why of their development. The story of solid propellants will have to be told by somebody else.

This is, in many ways, an auspicious moment for such a book. Liquid propellant research, active during the late 40's, the 50's, and the first half of the 60's, has tapered off to a trickle, and the time seems ripe for a summing up, while the people who did the work are still around to answer questions. Everyone whom I have asked for information has been more than cooperative, practically climbing into my lap and licking my face. I have been given reams of unofficial and quite priceless information, which would otherwise have perished with the memories of the givers. As one of them wrote to me, "What an opportunity to bring out repressed hostilities!" I agree.

My sources were many and various. Contractor and government agency progress (sometimes!) reports, published collections of papers presented at various meetings, the memories of participants in the story, intelligence reports; all have contributed. Since this is not a formal history, but an informal attempt by an active participant to tell the story as it happened, I haven't attempted formal documentation. Particularly as in many cases such documentation would be embarrassing—not to say hazardous! It's not only newsmen who have to protect their sources.

And, of course, I have drawn on my own records and recollections. For something more than twenty years, from 1 November 1949, when I joined the U.S. Naval Air Rocket Test Station, until 2 January 1970, when I retired from its successor, the Liquid Rocket Propulsion Laboratory of Picatinny Arsenal, I was a member of the unofficial, but very real, liquid propellant community, and was acutely aware of what was going on in the field, in this country and in England. (It wasn't until the late 50's that it was possible to learn much about the work in the Soviet Union, and propellant work outside these three countries has been negligible.)

The book is written not only for the interested layman—and for him I have tried to make things as simple as possible—but also for the professional engineer in the rocket business. For I have discovered that he is frequently abysmally ignorant of the history of his own profession, and, unless forcibly restrained, is almost certain to do something which, as we learned fifteen years ago, is not only stupid but is likely to result in catastrophe. Santayana knew exactly what he was talking about.

So I have described not only the brilliantly conceived programs of research and development, but have given equal time to those which, to put it mildly, were not so well advised. And I have told the stories of the triumphs of propellant research; and I have described the numerous blind alleys up which, from time to time, the propellant community unanimously charged, yapping as they went.

This book is opinionated. I have not hesitated to give my own opinion of a program, or of the intelligence—or lack of it—of the proposals made by various individuals. I make no apology for this, and can assure the reader that such criticism was not made with the advantage of 20–20 hindsight. At one point, in writing this book, when I had subjected one particular person's proposals to some rather caustic criticism, I wondered whether or not I had felt that way at the time they were made. Delving into my (very private) logbook, I found that I had described them then, simply as "Brainstorms and bullbleep!" So my opinion had not changed—at least, not noticeably.

I make no claim to completeness, but I have tried to give an accurate account of the main lines of research. If anyone thinks that I have unreasonably neglected his work, or doesn't remember things as I do, let him write to me, and the matter will be set right in the next (d.v.) edition. And if I seem to have placed undue emphasis on what happened in my own laboratory, it is not because my laboratory was unusual (although more nutty things seem to have happened there than in most labs) but that it was not, so that an account of what happened there is a good sample of the sort of things which were happening, simultaneously, in a dozen other laboratories around the country.

The treatment of individuals' names is, I know, inconsistent. The fact that the family name of somebody mentioned in the text is preceded by his given name rather than by his initials signifies only that I know him very well. Titles and degrees are generally ignored. Advanced degrees were a dime a dozen in the business. And the fact that an individual is identified in one chapter with one organization, and with another in the next, should be no cause for confusion. People in the business were always changing jobs. I think I set some sort of a record by staying with the same organization for twenty years.

One thing that is worth mentioning here is that this book is *about* a very few people. The propellant community—comprising those directing or engaged in liquid propellant research and development—was never large. It included, at the most, perhaps two hundred people, three-quarters of whom were serving merely as hands, and doing what the other quarter told them to do. That one quarter was a remarkably interesting and amusing group of people, including a surprisingly small number (compared to most other groups of the same size) of dopes or phoneys. We all knew each other, of course, which made for the informal dissemination of information at a velocity approaching that of light. I benefited particularly from this, since, as I was working for Uncle, and not for a rival contractor, nobody hesitated to give me "proprietory" information. If I wanted the straight dope from somebody, I knew I could get it at the bar at the next propellant meeting. (Many of the big propellant meetings were held in hotels, whose management, intelligently, would always set up a bar just outside the meeting hall. If the meeting wasn't in a hotel, I'd just look around for the nearest cocktail lounge; my man would probably be there.) I would sit down beside him, and, when my drink had arrived, ask, "Joe, what *did* happen on that last test firing you made? Sure, I've read your report, but I've written reports myself. What really happened?" Instant and accurate communication, without pain.

Conformists were hard to find in the group. Almost to a man, they were howling individualists. Sometimes they got along together—sometimes they didn't, and management had to take that into account. When Charlie Tait left Wyandotte, and Lou Rapp left Reaction Motors, and they both came to Aerojet, the management of the latter, with surprising intelligence, stationed one of them in Sacramento and one in Azusa, separated by most of the length of the state of California. Lou had been in the habit, when Charlie was giving a paper at a meeting, of slipping a nude or two into Charlie's collection of slides, and Charlie was no longer amused.

But friends or not, or feuding or not, everything we did was done with one eye on the rest of the group. Not only were we all intellectual rivals— "anything you can do I can do better"—but each of us knew that the others were the only people around competent to judge his work. Management seldom had the technical expertise, and since most of our work was classified, we couldn't publish it to the larger scientific community. So praise from the in-group was valued accordingly. (When Irv Glassman, presenting a paper, mentioned "Clark's classical work on explosive sensitivity," it put me on cloud nine for a week. *Classical*, yet!) The result was a sort of group Narcissism which was probably undesirable—but it made us work like Hell.

We did that anyway. We were in a new and exciting field, possibilities were unlimited, and the world was our oyster just waiting to be opened. We knew that we didn't have the answers to the problems in front of us, but we were sublimely confident of our ability to find them in a hurry, and set about the search with a "gusto"—the only word for it—that I have never seen before or since. I wouldn't have missed the experience for the world. So, to my dear friends and once deadly rivals, I say, "Gentlemen, I'm glad to have known you!"

John D. Clark
Newfoundland, N.J.
January 1971

Ignition!

1

How It Started

The dear Queen had finally gone to her reward, and King Edward VII was enjoying himself immensely as he reigned over the Empire upon which the sun never set. Kaiser Wilhelm II in Germany was building battleships and making indiscreet remarks, and in the United States President Theodore Roosevelt was making indiscreet remarks and building battleships. The year was 1903, and before its end the Wright brothers' first airplane was to stagger briefly into the air. And in his city of St. Petersburg, in the realm of the Czar of All the Russias, a journal whose name can be translated as "Scientific Review" published an article which attracted no attention whatsoever from anybody.

Its impressive but not very informative title was "Exploration of Space with Reactive Devices," and its author was one Konstantin Eduardovitch Tsiolkovsky, an obscure schoolteacher in the equally obscure town of Borovsk in Kaluga Province.

The substance of the article can be summarized in five simple statements.

1 Space travel is possible.
2 This can be accomplished by means of, and *only* by means of, rocket propulsion, since a rocket is the only known propulsive device which will work in empty space.
3 Gunpowder rockets cannot be used, since gunpowder (or smokeless powder either, for that matter) simply does not have enough energy to do the job.
4 Certain liquids *do* possess the necessary energy.
5 Liquid hydrogen would be a good fuel and liquid oxygen a good oxidizer, and the pair would make a nearly ideal propellant combination.

The first four of these statements might have been expected to raise a few eyebrows if anybody had been listening, but nobody was, and they were received with a deafening silence. The fifth statement was of another sort entirely, and a few years earlier would have been not merely surprising, but utterly meaningless. For liquid hydrogen and liquid oxygen were new things in the world.

Starting with Michael Faraday in 1823, scientists all over Europe had been trying to convert the various common gases to liquids—cooling them, compressing them, and combining the two processes. Chlorine was the first to succumb, followed by ammonia, carbon dioxide, and many others, and by the seventies only a few recalcitrants still stubbornly resisted liquefaction. These included oxygen, hydrogen and nitrogen (fluorine had not yet been isolated and the rare gases hadn't even been discovered), and the holdouts were pessimistically called the "permanent gases."

Until 1883. In April of that year, Z. F. Wroblewski, of the University of Krakow, in Austrian Poland, announced to the French Academy that he and his colleague K. S. Olszewski had succeeded in their efforts to liquefy oxygen. Liquid nitrogen came a few days later, and liquid air within two years. By 1891 liquid oxygen was available in experimental quantities, and by 1895 Linde had developed a practical, large-scale process for making liquid air, from which liquid oxygen (and liquid nitrogen) could be obtained, simply by fractional distillation.

James Dewar (later Sir James, and the inventor of the Dewar flask and hence of the thermos bottle), of the Royal Institute in London, in 1897 liquefied fluorine, which had been isolated by Moisson only eleven years before, and reported that the density of the liquid was 1.108. This wildly (and inexplicably) erroneous value (the actual density is 1.50) was duly embalmed in the literature, and remained there, unquestioned, for almost sixty years, to the confusion of practically everybody.

The last major holdout—hydrogen—finally succumbed to his efforts, and was liquefied in May of 1898. And, as he triumphantly reported, "on the thirteenth of June, 1901, five liters of it (liquid hydrogen) were successfully conveyed through the streets of London from the laboratory of the Royal Institution to the chambers of the Royal Society!"

And only *then* could Tsiolkovsky write of space travel in a rocket propelled by liquid hydrogen and liquid oxygen. Without Wroblewski and Dewar, Tsiolkovsky would have had nothing to talk about.

In later articles, Tsiolkovsky discussed other possible rocket fuels—methane, ethylene, benzene, methyl and ethyl alcohols, turpentine, gasoline, kerosene—practically everything that would pour and burn, but he apparently never considered any oxidizer other than liquid oxygen. And although he wrote incessantly until the day of his death (1935) his rockets remained on paper. He never *did* anything about them. The man who did was Robert H. Goddard.

As early as 1909 Dr. Goddard was thinking of liquid rockets, and came to the same conclusions as had his Russian predecessor (of whom he had never heard); that liquid hydrogen and liquid oxygen would be a near-ideal combination. In 1922, when he was Professor of Physics at Clark University, he started actual experimental work on liquid rockets and their components. Liquid hydrogen at that time was practically impossible to come by, so he worked with gasoline and liquid oxygen, a combination which he used in all of his subsequent experimental work. By November 1923 he had fired a rocket motor on the test stand, and on March 16, 1926, he achieved the first flight of a liquid-propelled rocket. It flew 184 feet in 2.5 seconds. (Exactly forty years later, to the day, Armstrong and Scott were struggling desperately to bring the wildly rolling Gemini 8 under control.)

One odd aspect of Goddard's early work with gasoline and oxygen is the very low oxidizer-to-fuel ratio that he employed. For every pound of gasoline he burned, he burned about 1.3 or 1.4 pounds of oxygen, when three pounds of oxygen would have been closer to the optimum. As a result, his motors performed very poorly, and seldom achieved a specific impulse of more than 170 seconds. (The specific impulse is a measure of performance of a rocket and its propellants. It is obtained by dividing the thrust of the rocket in pounds, say, by the consumption of propellants in pounds per second. For instance, if the thrust is 200 pounds and the propellant consumption is one pound per second, the specific impulse is 200 seconds.) It seems probable that he worked off-ratio to reduce the combustion temperature and prolong the life of his hardware—that is, simply to keep his motor from burning up.

The impetus for the next generation of experimenters came in 1923, from a book by a completely unknown Transylvanian German, one Herman Oberth. The title was *Die Rakete zu den Planetenraumen,* or *The Rocket into Planetary Space,* and it became, surprisingly, something of a minor best seller. People started thinking about rockets—practically nobody had heard of Goddard, who worked in exaggerated and unnecessary secrecy—and some of the people who thought about rockets decided to do something about them. First, they organized societies. The Verein fur Raumschiffart, or Society for Space Travel, generally known as the VfR, was the first, in June 1927. The American Interplanetary Society was founded early in 1930, the British Interplanetary Society in 1933, and two Russian groups, one in Leningrad and one in Moscow, in 1929. Then, they lectured and wrote books about rockets and interplanetary travel. Probably the most important of these was Robert Esnault-Pelterie's immensely detailed *L'Astronautique,* in 1930. And Fritz Lang made a movie about space travel—*Frau in Mond,* or *The Woman on the Moon,* and hired Oberth as technical adviser. And it was agreed that Lang and the film company (UFA) would put up the money necessary for Oberth to design and build a liquid-fueled rocket which would be fired, as a publicity stunt, on the day of the premiere of the movie.

The adventures of Oberth with the movie industry—and vice versa—are a notable contribution to the theater of the absurd (they have been described elsewhere, in hilarious detail), but they led to one interesting, if abortive, contribution to propellant technology. Foiled in his efforts to get a gasoline-oxygen rocket flying in time for the premiere of the movie (the time available was ridiculously short) Oberth designed a rocket which, he hoped, could be developed in a hurry. It consisted of a long vertical aluminum tube with several rods of carbon in the center, surrounded by liquid oxygen. The idea was that the carbon rods were to burn down from the top at the same rate as the oxygen was to be consumed, while the combustion gases were ejected through a set of nozzles at the top (forward) end of the rocket. He was never able to get it going, which was probably just as well, as it would infallibly have exploded. But—it was the first recorded design of a hybrid rocket—one with a solid fuel and a liquid oxidizer. (A "reverse" hybrid uses a solid oxidizer and a liquid fuel.)

At any rate, the premiere came off on October 15, 1929 (without rocket ascent), and the VfR (after paying a few bills) fell heir to Oberth's equipment, and could start work on their own in early 1930.

But here the story starts to get complicated. Unknown to the VfR—or to anybody else—at least three other groups were hard at work. F. A. Tsander, in Moscow, headed one of these. He was an aeronautical engineer who had written extensively—and imaginatively—on rockets and space travel, and in one of his publications had suggested that an astronaut might stretch his fuel supply by imitating Phileas Fogg. When a fuel tank was emptied, the astronaut could simply grind it up and add the powdered aluminum thus obtaining to the remaining fuel, whose heating value would be correspondingly enhanced! This updated emulation of the hero of *Around the World in Eighty Days,* who, when he ran out of coal, burned up part of his ship in order to keep the rest of it moving, not unnaturally remained on paper, and Tsander's experimental work was in a less imaginative vein. He started work in 1929, first with gasoline and gaseous air, and then, in 1931, with gasoline and liquid oxygen.

Another group was in Italy, headed by Luigi Crocco, and financed, reluctantly, by the Italian General Staff.*

Crocco started to work on liquid rockets in 1929, and by the early part of 1930 was ready for test firings. His work is notable not only for the surprising sophistication of his motor design, but above all for his propellants.

---

* The fact that the whole project was headed by a General G. A. Crocco is no coincidence. He was Luigi's father, and an Italian father is comparable to a Jewish mother.

He used gasoline for his fuel, which is not surprising, but for his oxidizer he broke away from oxygen, and used nitrogen tetroxide, N_2O_4. This was a big step—nitrogen tetroxide, unlike oxygen, can be stored indefinitely at room temperature—but nobody outside of his own small group heard of the work for twenty-four years![*]

V. P. Glushko, another aeronautical engineer, headed the rocket group in Leningrad. He had suggested suspensions of powdered beryllium in oil or gasoline as fuels, but in his first firings in 1930, he used straight toluene. And he took the same step—independently—as had Crocco. He used nitrogen tetroxide for his oxidizer.

The VfR was completely unaware of all of this when they started work. Oberth had originally wanted to use methane as fuel, but as it was hard to come by in Berlin, their first work was with gasoline and oxygen. Johannes Winkler, however, picked up the idea, and working independently of the VfR, was able to fire a liquid oxygen-liquid methane motor before the end of 1930. This work led nowhere in particular, since, as methane has a performance only slightly superior to that of gasoline, and is much harder to handle, nobody could see any point to following it up.

Much more important were the experiments of Friedrich Wilhelm Sander, a pyrotechnician by trade (he made commercial gunpowder rockets) who fired a motor early in March 1931. He was somewhat coy about his fuel, calling it merely a "carbon carrier," but Willy Ley has suggested that it may well have been a light fuel oil, or benzene, into which had been stirred considerable quantities of powdered carbon or lampblack. As a pyrotechnician, Sander would naturally think of carbon as *the* fuel, and one Hermann Noordung (the pseudonym of Captain Potocnik of the old Imperial Austrian army), the year before, had suggested a suspension of carbon in benzene as a fuel. (The idea

---

[*] In a letter to *El Comercio,* of Lima, Peru, 7 October, 1927, one Pedro A. Paulet, a Peruyian chemical engineer, claimed to have experimented—in 1895–97 (!)—with a rocket motor burning gasoline and nitrogen tetroxide. If this claim has any foundation in fact, Paulet anticipated not only Goddard but even Tsiolkovsky.

However, consider these facts. Paulet claimed that his motor produced a thrust of 200 pounds, and that it fired intermittently, 300 times a minute, instead of continuously as conventional rocket motors do.

He also claimed that he did his experimental work *in* Paris.

Now, I know how much noise a 200-pound motor makes. And I know that if one were fired three hundred times a minute—the rate at which a watch ticks—it would sound like a whole battery of fully automatic 75 millimeter antiaircraft guns. Such a racket would have convinced the Parisians that the Commune had returned to take its vengeance on the Republic, and would certainly be remembered by *somebody* beside Paulet! But only Paulet remembered.

In my book, Paulet's claims are completely false, and his alleged firings never took place.

was to increase the density of the fuel, so that smaller tanks might be used.) The important thing about Sander's work is that he introduced another oxidizer, red fuming nitric acid. (This is nitric acid containing considerable quantities—5 to 20 or so percent—of dissolved nitrogen tetroxide.) His experiments were the start of one of the main lines of propellant development.

Esnault-Pelterie, an aviation pioneer and aeronautical engineer, during 1931, worked first with gasoline and oxygen, and then with benzene and nitrogen tetroxide, being the third experimenter to come up, independently, with this oxidizer. But that was to be a repeating pattern in propellant research—half a dozen experimenters generally surface simultaneously with identical bones in their teeth! His use of benzene (as Glushko's of toluene) as a fuel is rather odd. Neither of them is any improvement on gasoline as far as performance goes, and they are both much more expensive. And then Esnault-Pelterie tried to use tetranitromethane, $C(NO_2)_4$ for his oxidizer, and promptly blew off four fingers. (This event was to prove typical of TNM work.)

Glushko in Leningrad took up where Sander had left off, and from 1932 to 1937 worked with nitric acid and kerosene, with great success. The combination is still used in the USSR. And in 1937, in spite of Esnault-Pelterie's experience, which was widely known, he successfully fired kerosene and tetranitromethane. This work, however, was not followed up.

Late in 1931 Klaus Riedel of the VfR designed a motor for a new combination, and it was fired early in 1932. It used liquid oxygen, as usual, but the fuel, conceived by Riedel and Willy Ley, was a 60–40 mixture of ethyl alcohol and water. The performance was somewhat below that of gasoline, but the flame temperature was much lower, cooling was simpler, and the hardware lasted longer. This was the VfR's major contribution to propellant technology, leading in a straight line to the A-4 (or V-2) and it was its last. Wernher von Braun started work on his PhD thesis on rocket combustion phenomena at Kummersdorf-West in November 1932 under Army sponsorship, the Gestapo moved in on the rest of the VfR, and the society was dead by the end of 1933.

Dr. Eugen Sänger, at the University of Vienna, made a long series of firings during 1931 and 1932. His propellants were conventional enough—liquid (or sometimes gaseous) oxygen and a light fuel oil—but he introduced an ingenious chemical wrinkle to get his motor firing. He filled the part of his fuel line next to the motor with diethyl zinc, to act as what we now call a "hypergolic starting slug." When this was injected into the motor and hit the oxygen it ignited spontaneously, so that when the fuel oil arrived the fire was already burning nicely. He also compiled a long list, the first of many, of possible fuels, ranging from hydrogen to pure carbon, and calculated the performance of each with oxygen and with N_2O_5. (The latter, being not only unstable, but a solid to boot, has naturally never been used.) Unfortunately, in his calculations he somewhat naively assumed 100 percent thermal efficiency, which would involve either (a) an infinite chamber pressure, or (b) a zero exhaust pressure firing into a perfect vacuum, and in either case would require an infinitely long nozzle, which might involve some difficulties in fabrication. (Thermal efficiencies in a rocket usually run around 50 or 60 percent.) He also suggested that ozone might be used as an oxidizer, and as had Tsander, that powdered aluminum might be added to the fuel.

Then Luigi Crocco, in Italy, had another idea, and was able to talk the Ministry of Aviation into putting up a bit of money to try it out. The idea was that of a monopropellant. A monopropellant is a liquid which contains in itself both the fuel and the oxidizer, either as a single molecule such as methyl nitrate, CH_3NO_3 in which the oxygens can burn the carbon and the hydrogens, or as a mixture of a fuel and an oxidizer, such as a solution of benzene in N_2O_4. On paper, the idea looks attractive. You have only one fluid to inject into the chamber, which simplifies your plumbing, your mixture ratio is built in and stays where you want it, you don't have to worry about building an injector which will mix the fuel and the oxidizer properly, and things are simpler all around. *But!* Any intimate mixture of a fuel and an oxidizer is a potential explosive, and a molecule with one reducing (fuel) end and one oxidizing end, separated by a pair of firmly crossed fingers, is an invitation to disaster.

All of which Crocco knew. But with a species of courage which can be distinguished only with difficulty from certifiable lunacy, he started in 1932 on a long series of test firings with nitroglycerine (no less!) only sightly tranquilized by the addition of 30 percent of methyl alcohol. By some miracle he managed to avoid killing himself, and he extended the work to the somewhat less sensitive nitromethane, CH_3NO_2. His results were promising, but the money ran out in 1935, and nothing much came of the investigation.

Another early monopropellant investigator was Harry W. Bull, who worked on his own at the University of Syracuse. By the middle of 1932 he had used gaseous oxygen to burn gasoline, ether, kerosene, fuel oil, and alcohol. Later he tried, without success, to burn alcohol with 30 percent hydrogen peroxide (the highest strength available in the U.S. at the time), and to burn turpentine with (probably 70 percent) nitric acid. Then, in 1934 he tried a monopropellant of his own invention, which he called "Atalene," but did not otherwise identify. It exploded and put him in the hospital. Dead end.

And Helmuth Walter, at the Chemical State Institute in Berlin, in 1934 and 1935 developed a monopropellant motor which fired 80 percent hydrogen peroxide, which had only lately become available. When suitably catalyzed, or when heated, hydrogen peroxide decomposes into oxygen and superheated steam, and thus can be used as a monopropellant. This work was not made public—the Luftwaffe could see uses for it—but it was continued and led to many things in the next few years.

The last strictly prewar work that should be considered is that of Frank Malina's group at GALCIT. (Guggenheim Aeronautical Laboratories, California Institute of Technology.) In February of 1936 he planned his PhD thesis project, which was to be the development of a liquid-fueled sounding rocket. The group that was to do the job was gradually assembled, and was complete by the summer of 1937: six people, included Malina himself, John W. Parsons, the chemist of the group, Weld Arnold, who put up a little money, and Hsu Shen Tsien, who, thirty years later, was to win fame as the creator of Communist China's ballistic missiles. The benign eye of Theodore von Kármán watched over the whole.

The first thing to do was to learn how to run a liquid rocket motor, and experimental firings, with that object in view, started in October 1936. Methanol and gaseous oxygen were the propellants. But other propellants were considered, and by June 1937, Parsons had compiled lists, and calculated the performances (assuming, as had Sänger, 100 percent efficiency) of dozens of propellant combinations. In addition to Sänger's fuels, he listed various alcohols and saturated and unsaturated hydrocarbons, and such exotic items as lithium methoxide, dekaborane, lithium hydride, and aluminum triemethyl. He listed oxygen, red fuming nitric acid, and nitrogen tetroxide as oxidizers.

The next combination that the group tried then, was nitrogen tetroxide and methanol. Tests began in August 1937. But Malina, instead of working outdoors, as any sane man would have done, was so ill advised as to conduct his tests in the Mechanical Engineering building, which, on the occasion of a misfire, was filled with a mixture of methanol and N_2O_4 fumes. The latter, reacting with the oxygen and the moisture in the air, cleverly converted itself to nitric acid, which settled corrosively on all the expensive machinery in the building. Malina's popularity with the establishment suffered a vertiginous drop, he and his apparatus and his accomplices were summarily thrown out of the building, and he was thereafter known as the head of the "suicide squad." Pioneers are seldom appreciated.

But the group continued work, until July 1, 1939, when, at the instigation of General Hap Arnold, the Army Air Corps sponsored a project to develop a JATO—a rocket unit to help heavily laden planes take off from short runways.

From now on, rocket research was to be paid for by the military, and was to be classified. GALCIT had lost her virginity with Malina's first explosion. Now she had lost her amateur standing.

2

Peenemunde
and JPL

Von Braun started work on his PhD thesis (rocket combustion processes) in November 1932. All of his experimental work was done at Kummersdorf-West, an artillery range near Berlin—and the Reichswehr paid the freight, and built up a rocket establishment around him. When he got his degree, in 1937, he was made the technical director of the organization, which was soon moved to Peenemunde. There the A-4, better known by its propaganda name "V-2" was designed and developed.

Very little propellant development was involved in the A-4. From the beginning, liquid oxygen was the intended oxidizer, and 70–30 alcohol-water mixture (as had been used by the VfR) the fuel. And Helmuth Walter's 80 percent hydrogen peroxide was used to drive the fuel pumps. The peroxide entered a decomposition chamber, where it was mixed with a small quantity of a solution of calcium permanganate in water. This catalyzed its decomposition into oxygen and superheated steam, which drove the turbines which drove the pumps which forced the oxygen and the alcohol into the main combustion chamber.

The A-4 was a long range strategic weapon, not designed to be fired at a moment's notice. It was perfectly practical to set it up, and then load it with alcohol and oxygen just before firing. But the Reichswehr needed antiaircraft rockets that were always ready to fire. When you get word from your forward observers that the bombers are on the way, you don't have time to load up a missile with liquid oxygen. What you need is a storable propellant—one that can be loaded into the tanks beforehand—and kept there until you push the button. You can't do that with oxygen, which cannot be kept liquid above −119°C, its critical temperature, by any pressure whatsoever.

The Reichswehr was rather slow to realize the need for AA rockets—maybe they believed Hermann Goering when he boasted, "If the British ever bomb Berlin, you can call me Meyer!"—but when they did they found that work on storable propellants was well under way. It was, at first, concentrated at Helmuth Walter's Witte Werke at Kiel. As has been mentioned, high strength hydrogen peroxide (80–83 percent) first became available in about 1934, and Walter had fired it as a monopropellant, and the Luftwaffe was immensely interested. Like General Arnold, in the U.S. they could appreciate the fact that a JATO rocket would enable a bomber to take off with a heavier load than it could normally carry, and by February 1937, a Walter hydrogen peroxide JATO had helped a Heinkel Kadett airplane to get off the ground. Later in the year, a rocket powered airplane was flown—again using a hydrogen peroxide motor. The Messerschmitt 163-A interceptor used the same propellant.

But peroxide is not only a monopropellant, it's also a pretty good oxidizer. And Walter worked out a fuel for it that he called "C-Stoff." (The peroxide itself was called "T-Stoff.") Hydrazine hydrate, $N_2H_4 \cdot H_2O$ ignited spontaneously when it came in contact with peroxide (Walter was probably the first propellant man to discover such a phenomenon) and C-Stoff consisted of 30 percent hydrazine hydrate, 57 of methanol, and 13 of water, plus thirty milligrams per liter of copper as potassium cuprocyanide, to act as an ignition and combustion catalyst. The reason for the methanol and the water was the fact that hydrazine hydrate was hard to come by—so hard, in fact, that by the end of the war its percentage in C-Stoff was down to fifteen. The Messerschmitt 163-B interceptor used C-Stoff and T-Stoff.

The next organization to get into the rocket business was the Aeronautical Research Institute at Braunschweig. There, in 1937–38, Dr. Otto Lutz and Dr. Wolfgang C. Noeggerath started to work on the C-Stoff-T-Stoff combination. Next, BMW (Bavarian Motor Works—yes, the people who make the motorcycles) were invited by the Luftwaffe to get into the act. Helmut Philip von Zborowski, the nephew of the famous pre-World War I racing driver, was in charge of the operation, and Heinz Mueller was his second. In the summer of 1939 BMW got a contract to develop a JATO unit, using the C-T-Stoff combination, and they worked with it for some months. But von Zborowski was convinced that 98 percent nitric acid was the better oxidizer, as well as being immensely easier to get (I.G. Farben guaranteed unlimited quantities), and set out to convert the brass to his point of view. From the beginning of 1940, he and Mueller worked on the nitric acid–methanol combination, and in 1941 proved his point, convincingly, with a perfect thirty-second run at the three thousand pounds force thrust level. He even convinced Eugen Sänger, who was sure that oxygen was the only oxidizer worth thinking about.

And in the meantime, early in 1940, he and Mueller had made an immensely important discovery—that certain fuels (aniline and turpentine were the first they found) ignited spontaneously upon contact with nitric acid. Noeggerath learned of this, and joined the BMW people in their search for fuels with this interesting property. His code name for nitric acid was "Ignol" and for his fuels "Ergol," and, a fast man with a Greek root, he came up with "Hypergol" for the spontaneous igniters. "Hypergol" and its derivatives, such as the adjective "hypergolic" have become a permanent part not only of the German, but of the English language, and even, in spite of the efforts of Charles de Gaulle to keep the language "pure," of the French as well.

The discovery of hypergolicity was of major importance. Running a rocket motor is relatively easy. Shutting it down without blowing something up is harder. But starting it up without disaster is a real problem. Sometimes electrical igniters are used—sometimes pyrotechnic devices. But neither can always be trusted, and either is a nuisance, an added complication, when you already have more complications than you want. Obviously, if your combination is hypergolic, you can throw out all the ignition schemes and devices, and let the chemistry do the work. The whole business is much simpler and more reliable.

But as usual, there's a catch. If your propellants flow into the chamber and ignite immediately, you're in business. But if they flow in, collect in a puddle, and *then* ignite, you have an explosion which generally demolishes the engine and its immediate surroundings. The accepted euphemism for this sequence of events is a "hard start." Thus, a hypergolic combustion must be *very* fast, or it is worse than useless. The Germans set an upper limit of 50 milliseconds on the ignition delay that they could tolerate.

Incidentally, and to keep the record straight, Zborowski named *his* propellants after plants. Nitric acid he called "Salbei" for sage, and his fuels "Tonka," after the bean from which coumarin, which smells like vanilla, is extracted. Considering the odors of the things he worked with, I can't think of more inappropriate names!

The first ignition delay tests were, to put it mildly, somewhat primitive. After a long night session, searching through old chemistry texts for substances that were violently reactive with nitric acid, Zborowski and Mueller would soak a wiping rag with a promising candidate and spray it with nitric acid and see how quickly—or if—it burst into flames. And they ran into a peculiar phenomenon. An old, used wiping rag from the machine shop would sometimes ignite much faster than a new clean one soaked with the same fuel. Their chemistry laboratory furnished them with the answer. Traces of iron and copper from the shop, as the metals or as salts, catalyzed the ignition reaction. So they modified their 98 percent nitric acid, "Salbei" by adding to it 6 percent of hydrated ferric chloride, and called the new oxidizer "Salbeik."

The wiping-rag technique was soon supplanted by a somewhat more sophisticated gadget with which you could drop a single drop of a candidate fuel into a thimbleful of acid, and determine its hypergolic properties with less risk of setting fire to the whole shop, and for the next four years BMW on the one hand and Noeggerath on the other were trying the hypergolicity of everything they could lay their hands on. At BMW, where propellant development was directed by Hermann Hemesath, more than 2000 prospective fuels were tried. And very soon the I.G. Farben organization at Ludwigshaven started doing the same thing. With a deplorable lack of imagination, Farben eschewed code names at first, and labeled their mixtures with code numbers like T93/4411.

The fuels that the three organizations developed were many and various, but at the same time very much alike, since there was a limited number of compounds which were hypergolic with nitric acid—and available in any quantity. Tertiary amines, such as triethyl amine were hypergolic, and aromatic amines, such as aniline, toluidine, xylidine, N-methylaniline were even more so. Most of the mixtures tried—neat fuels consisting of a single pure compound were unheard of—were based on the aniline family, frequently with the addition of triethylamine, plus, at times, things like xylene, benzene, gasoline, tetrahydrofuran, pyrocatechol, and occasionally other aliphatic amines. The BMW Tonka 250 comprised 57 percent of raw xylidine and 43 of triethylamine (it was used in the "Taifun" missile) and Tonka 500 contained toluidine, triethylamine, aniline, gasoline, benzene, and raw xylidine. Noeggerath added furfuryl alcohol to Tonka 250 to get "Ergol-60" which he considered the "best" hypergol, and reported, somewhat wistfully, that furfuryl alcohol was readily available in the United States—as it was not in Germany.

As soon as one of the investigators found a mixture that he liked he applied for a patent on it. (Such an application would probably not even be considered under the much stricter U.S. patent laws.) Not surprisingly, everybody and Hemesath and Noeggerath in particular, was soon accusing everybody else of stealing his patent. In 1946, when Heinz Mueller came to this country, he met Noeggerath again, and found him still indignant, bursting out with "And BMW, especially Hemesath, did *swipe* a lot of patents from us!"

Around 1942 or 1943 I.G. Farben shifted the emphasis of their fuel work away from the mixtures they had been working with at first, and which were so similar to the Tonkas and the Ergols, to a series of fuels based on the "Visols," which were vinyl ethers. The vinyl ethers were very rapidly hypergolic with MS-10, a mixed acid consisting of 10 percent sulfuric acid and 90 percent nitric, and the ignition delay was less sensitive to temperature than it was with straight nitric. (This had been a serious problem. A propellant pair might ignite in 50 milliseconds at room temperature, and wait around a whole second at 40 below.) Also, it was believed, practically as an article of faith, that

MS-10 did not corrode stainless steel. This was a delusion that lasted five years before it was punctured.

A typical mixture, patented by Dr. Heller in 1943, consisted of 57.5 percent Visol-1 (vinylbutyl ether) or Visol-6 (vinylethyl ether) 25.8 percent Visol-4 (divinylbutanediolether) 15 percent aniline, and 1.7 percent of iron penta-carbonyl or iron naphthenate. (Heller had to put his iron catalyst in his fuel rather than in his oxidizer, since the latter contained sulfuric acid, and iron sulfates are insoluble in nitric acid.) There were many variations on these fuels, vinylisobutyl ether being substituted at times for the n-butyl compound. All in all, more than 200 mixtures were tried, of which less than ten were found satisfactory. "Optolin" was a mixture of aniline, a Visol, aromatics, sometimes amines, gasoline, and pyrocatechol. The Wasserfall SAM used a Visol fuel.

Several agencies tried to discover additives which, in small quantities, would make gasoline or benzene or methanol hypergolic with acid. Things like iron carbonyl and sodium selenide were more or less successful, but the success was academic at best, since the useful additives were all either too rare, too expensive, or too active to live with.

But nitric acid was definitely the winner. Many German missiles were designed, at first, to use peroxide, but as the war went on, the Walter Type XVII submarines threatened to use up the whole production, and as the nitric acid work was so successful, the shift to the latter oxidizer for missile work was inevitable. During this period many other combinations than those actually tried were considered, and theoretical performances were calculated. These calculations were not the early naïve estimates of Sänger et al., but considered the combustion pressure, the exhaust pressure, thermal efficiency, temperature of combustion, dissociation—the whole business. Such exact calculations are outrageously tedious—a single one done with a desk calculator, can easily take a whole day. But Dr. Grete Range and others struggled through them, considering as fuels, alcohol, alcohol-water, gasoline, diesel fuel, ammonia, propargyl alcohol, and God only knows what else, and as oxidizers, oxygen, nitric acid, N_2O_4, tetranitromethane, ozone, and OF_2, although the laboratory men were never able to lay their hands on enough of the last to characterize it. And as early as 1943 they were thinking of using chlorine trifluoride, which before that had been nothing but a laboratory curiosity. But it had recently been put into production—its intended use was an incendiary agent—and they calculated its performance too, with ammonia and with such oddities as a suspension of carbon in water.

One calculation made at this time by Dr. Noeggerath, showed that if the propellants in the A-4 were replaced by nitric acid and diesel fuel, the range of the missile would be increased by an appreciable percentage—not because their propellants had a better performance than the oxygen-alcohol combination actually used, which they did not, but because their higher density allowed more propellant to be stuffed into the tanks. This calculation had no particular effect at that time, although the A-10, a planned successor to the A-4, was to have used the new combination, but some years later, in Russia, the consequences were to be hilarious.

The oxidizer that was always a "might have been" was tetranitromethane. It's a good oxidizer, with several advantages. It's storable, has a better performance than nitric acid, and has a rather high density, so you can get a lot of it in a small tank. But it melts at +14.1°C so that at any time other than a balmy summer day it's frozen solid. And it can explode—as Esnault-Pelterie had discovered, and it took out at least one German laboratory. The eutectic mixture with N_2O_4, 64 percent TNM, 36 N_2O_4, doesn't freeze above −30°C, and is considerably less touchy than is straight TNM, but it was still considered dangerous, and Noeggerath refused to have anything to do with it or, even to permit it in his laboratory. But the engineers kept looking at it wistfully, and when they received a (completely false) intelligence report that it was being used on a large scale in the United States, the Germans heroically started synthesis, and had accumulated some eight or ten tons of the stuff by the end of the war. Nobody ever found any use for it.

Another idea which didn't get anywhere, was that of a heterogeneous fuel—a suspension, or slurry, of a powdered metal, such as aluminum, in a liquid fuel such as gasoline. This had been suggested by several writers, among them Tsander in Russia and Sänger in Austria, and Heinz Mueller of BMW tried it out, using powdered aluminum or magnesium in diesel oil. The performance was very poor—the chamber pressure was 50 to 100 psi instead of the 300 they were shooting for—due to the incomplete combustion of the metal. But the other results were spectacular. The motor was fired in a horizontal position against an inclined wall to deflect the exhaust stream upwards. But the unburned metal particles settled down and decorated all the pine trees in the vicinity with a nice, shiny, silvery coating—very suitable for Christmas trees. The slurry idea was to emerge again twenty years later, to drive another generation of experimenters crazy.

Experimentation on monopropellants (which were called "Monergols") continued until the end of the war. In 1937–1938 a good deal of work was attempted with solutions of N_2O or NH_4NO_3 in ammonia. (The latter mixture, under the name of Driver's solution, had been known for many years.) The only result of these experiments was a depressing series of explosions and demolished motors. And at Peenemunde, a Dr. Wahrmke tried dissolving alcohol in 80 percent H_2O_2 and then firing *that* in a motor. It detonated, and killed him. The Wm. Schmidding firm, nevertheless, kept on experimenting with a monopropellant they called "Myrol," an 80–20 mixture of methyl nitrate and methanol—very similar to the nitroglycerine-methanol mixture that Crocco had tried years before. They managed to fire the material, and got a fairly respectable performance, but they were plagued by explosion after explosion, and were never able to make the system reliable.

And there was finally the propellant combination that the BMW people and those at ARIB called the "Lithergols"—which was really a throwback to the original hybrid motor tried by Oberth during the UFA period. Peroxide or nitrous oxide, N_2O, was injected into a motor in which several sticks of porous carbon were secured. Nitrous oxide can decompose exothermically into oxygen and nitrogen, as peroxide does to oxygen and steam, and can thus act as a monopropellant, but the experimenters wanted to get extra energy from the combustion of the carbon by the oxygen formed. When they surrendered to the Americans at the end of the war, they assured their captors that just a little more engineering work was needed to make the system work properly. Actually some twenty years elapsed before anybody could make a hybrid work.

Meanwhile, back at the ranch—

The most striking thing about propellant research in the United States during the war years is how closely it paralleled that in Germany. True, there was no American A-4, and high strength hydrogen peroxide was unobtainable in this country, but the other developments were closely similar.

As mentioned in the first chapter, GALCIT's first job for the armed forces was to produce a JATO to help the Army Air Corps get its bombers off the ground. And the Air Corps demanded a storable oxidizer—they were not, repeat not, going to fool around with liquid oxygen.

So the first order of business was choosing an oxidizer. Oxygen and ozone, neither of them storable, were obviously out. Chlorine had insufficient energy, and Malina, Parsons, and Forman who, with the assistance of Dr. H. R. Moody, did a survey of the subject, considered that N_2O_4 was impractical. It is difficult to say why, but the extremely poisonous nature of the beast may have had something to do with its rejection. They considered 76 percent perchloric acid, and tetranitromethane, and finally settled on red fuming nitric acid, RFNA, containing 6 or 7 percent N_2O_4. They tried crucible burning of various fuels with this acid—gasoline, petroleum ether, kerosene, methyl and ethyl alcohol, turpentine, linseed oil, benzene, and so on, and found that the acid would support combustion. Further, they found that hydrazine hydrate and benzene were hypergolic with it, although they had never heard of the word, so acid it was. There is a highly nonprophetic statement in the final Report for 1939–1940, Air Corps Jet Propulsion Research, GALCIT-JPL Report No. 3, 1940. (By now Malina's group had become the Jet Propulsion Laboratory, with von Kármán at the head.)

"The only possible source of trouble connected with the acid is its corrosive nature, which can be overcome by the use of corrosion-resistant materials."

Ha! If they had known the trouble that nitric acid was to cause before it was finally domesticated, the authors would probably have stepped out of the lab and shot themselves.

Be that as it may, the report was an excellent survey of the field as it was at that time, and contained sophisticated and accurate performance calculations. The procedure had been developed in Malina's 1940 PhD thesis, and was essentially and inevitably the same as that developed in Germany. One of the first compilations of the thermodynamic properties of exhaust gases was published by J. O. Hirschfelder in November 1942, as necessary raw data for such computations.

Malina and company started experimental work with RFNA and gasoline as early as 1941—and immediately ran into trouble. This is an extraordinarily recalcitrant combination, beautifully designed to drive any experimenter out of his mind. In the first place, it's almost impossible to get it started. JPL was using a spark plug for ignition, and more often than not, getting an explosion rather than the smooth start that they were looking for. And when they did get it going, the motor would cough, chug, scream and hiccup—and then usually blow anyway. Metallic sodium suspended in the fuel helped the ignition somewhat, and benzene was a little better than gasoline—but not much, or enough. It took an accidental discovery from the other side of the country to solve their immediate problems.

Here we must backtrack. From 1936 to 1939, Robert C. Truax, then a midshipman at the U.S. Naval Academy, had been experimenting with liquid fueled rockets, on his own time and with scrounged material. He graduated, spent the required two years on sea duty, and in 1941, then a lieutenant commander, was ordered to the Engineering Experiment Station at Annapolis, with orders to develop a JATO. For the Navy was having trouble getting their underpowered and overloaded PBM and PBY patrol bombers off the water. And he, too, ran into ignition and combustion difficulties. But one of his small staff, Ensign Stiff, while working on gas generators (small combustion devices designed to supply hot gas under pressure) discovered that aniline and RFNA ignited automatically upon contact. (Such discoveries are usually surprising, not to say disconcerting, and one wonders whether or not Ensign Stiff retained his eyebrows.)

At any rate, Frank Malina, visiting EES in February of 1942, learned of this discovery, and instantly phoned JPL in Pasadena; and JPL immediately switched from gasoline to aniline. And their immediate difficulties miraculously disappeared. Ignition was spontaneous and immediate, and combustion was smooth. They had a 1000-pound thrust motor running by the first of April (these people were professionals by that time) and on the fifteenth it boosted an A20-A medium bomber into the air—the first flight of a liquid JATO in the United States.

Truax, of course, adopted the propellant combination, and early in 1943, hanging two 1500 pound units on a PBY, managed to get the much overloaded Dumbo off the water.

Other people were working on JATO's for the Navy, among them Professor Goddard himself, whose unit was successfully flown in a PBY in September 1942—the first Navy JATO. He used his classic combination of liquid oxygen and gasoline, but Reaction Motors, also active in the field, came up with an ingenious variation.

Reaction Motors, Inc., generally called RMI, was founded in 1941 by a handful of veterans of the American Rocket Society including James Wyld, Lovell Lawrence, and John Shesta, and undertook to build a JATO unit. They first used liquid oxygen—all the ARS work had been with that oxidizer—and gasoline. But they found that the combination was too hot, and burned out their motors. So, as the gasoline entered the chamber, they mixed it with water through a metering valve. Combustion was smoother, and the motor stayed in one piece. This was a somewhat less elegant solution to the problem of combustion temperatures than was that used by the VfR (and Peenemunde) when they mixed water with their alcohol fuel. The RMI unit was successfully flown in the PBM in 1943. During the trials, run on the Severn River, the exhaust jet set the tail of the seaplane on fire, but the test pilot rose (or sank) to the occasion and set the plane down, tail first on the water in the manner of an old time movie comedian with his coattails on fire, seating himself hurriedly in a washtub full of water, with appropriate hissing noises and clouds of steam.

The aniline-RFNA combination had the one—but magnificent—virtue that it worked. Otherwise it was an abomination. In the first place, aniline is much harder to come by than gasoline—particularly in the midst of a dress-shirt war, when everybody and his brother wants to use it for explosives and what not. Second, it is extremely poisonous, and is rapidly absorbed through the skin. And third, it freezes at $-6.2°C$, and hence is strictly a warm-weather fuel. The Army and the Navy both, in a rare example of unanimity, screamed at the thought of using it. But they had no choice.

Two closely interwound lines of research characterize the rest of the war period. One was designed to reduce the freezing point of aniline, the other was to make gasoline, somehow, hypergolic with nitric acid. American Cyanamid was given a contract to investigate additives which might have the latter effect and JPL worked both sides of the street, as well as experimenting with changes in the composition of the acid. Besides their usual RFNA, containing about 6 percent N_2O_4, they experimented with one containing about 13 percent, as well as with a mixed acid rather similar to that the Germans were using, but a little more potent. One mixture they used contained 88 percent nitric acid, 9.6 percent sulfuric, and 2.4 percent SO_3. (This was very similar to the mixed acids used in explosives manufacture.) And they, too, believed that it didn't corrode stainless steel.

The obvious way to lower the freezing point of aniline is to mix it with something else—preferably something that is as hypergolic as the aniline itself. And the obvious way to make gasoline hypergolic is to mix *it* with something that is. Both lines of endeavor were pursued with enthusiasm.

At LPL they mixed aniline with orthotoluide, its near relative, and got a eutectic freezing at −32°C. But o-toluidine was as scarce as aniline, and although the mixture was successfully fired, it never became operational. A more practical additive was furfuryl alcohol, for which Zborowski was pining. Furfuryl alcohol comes from oat hulls and Quaker Oats had tank cars of the stuff, which they were delighted to sell to anybody who would take it off their hands. And 20 percent of furfuryl alcohol in aniline reduced the freezing point to 0°F, or −17.8°C, and the eutectic mixture, 51 percent aniline, 49 furfuryl alcohol, had a freezing point of −42°C. And furfuryl alcohol itself was about as hypergolic as aniline.

And to gasoline, JPL added aniline, diphenylamine, mixed xylidines and other relatives of aniline; assorted aliphatic amines, and everything else they could think of, and then measured the ignition delay. But they never found an additive which, in small percentages, would make gasoline rapidly hypergolic, with either RFNA or mixed acid. One of their best additives was mixed xylidines, but it took about 50 percent of the xylidines in the mixture to make it reliably and rapidly hypergolic—which took it out of the additives class, and made it a major component. To make it more discouraging, there were no production facilities for the xylidines in the United States, and although Aerojet looked at a similar mixture a few years later (in 1949) it never came to anything.

American Cyanamid was having a similar experience. They started with #2 fuel oil, diesel oil, and gasoline, and added to the particular fuel aniline, dimethylaniline, mono- and diethylaniline, crude monoethylaniline—and turpentine. Most of their work was done with mixed acid, a little with RFNA, and some with straight 98 percent nitric acid (White Fuming Nitric Acid, or WFNA). And in no case did they find an effective additive. But they found that turpentine was magnificently hypergolic with mixed acid or RFNA, and might well be a good fuel all by itself. (And think of all those lovely votes from the piney woods of the South!)

Aerojet Engineering was founded in March of 1942, to act, essentially, as the manufacturing arm of JPL. The founders were von Kármán, Malina, Parsons, Summerfield, and Forman, all of JPL, plus Andrew Haley, who was von Kármán's attorney. And they started their own propellant research program, although for some years it was difficult to disentangle it from JPL's.

Aerojet was the first organization to work extensively with crude N-ethyl aniline, sometimes called monoethylaniline, as a fuel. This is almost as rapidly hypergolic as aniline. The crude or commercial product contains about 10 percent diethylaniline and 26 straight aniline, the remainder being the monoethyl compound, and its freezing point is about −63°C. All in all, it was an elegant answer to the freezing point problem, but it was just about as poisonous as its ancestor, and just as hard to come by.

But it could be lived with. The propellants for the Aerojet JATO, in production by the end of the war, were mixed acid and monoethylaniline, as were those of RMI's motor for the Navy's surface-to-air missile, Lark, whose development started in 1944. The surface-to-surface Corporal, started the same year, was designed around the RFNA-aniline-furfuryl alcohol combination.

Three organizations worked on monopropellants during the war although the effort was limited. All of them concentrated on nitromethane. JPL worked on it first, in 1944, or earlier, and found that its combustion was improved by the addition of small quantities of chromium trioxide (later chromium acetyl-acetonate) to the fuel. Aero-jet also worked with it, and found that it was necessary to desensitize it by the addition of 8 percent of butyl alcohol. And Bob Truax, at KES, tried his hand—and was almost killed when somebody connected the wrong pipe to the right valve and the tank blew. And finally Dave Altman, at JPL, tried a mixture of benzene and tetranitromethane, which naturally detonated at once.

And then the war was over, and the German work came to light—and things started to get really complicated.

3

The Hunting of
the Hypergol . . .

As the American interrogators moved into Germany close behind—and sometimes ahead of—the armies, they found the German rocket scientists more than willing to surrender (and get new jobs) and more than anxious to tell everything they knew. Not only did the Americans get almost all the top scientists—they got everything else that wasn't nailed down, including the complete Peenemunde archives (which von Braun's crew had thoughtfully deposited in an abandoned mine) and all the A-4 rockets, complete or otherwise. And, red-blooded young Americans all, with larceny in their hearts, they liberated every milligram of hydrazine hydrate and high-strength hydrogen peroxide that they could find in Germany. Plus, naturally, the special aluminum tank cars built to carry the latter. Everything was promptly shipped to the United States.

These steps were obvious. The next step was not.

The alcohol–oxygen combination seemed all right for long-range missiles, but the United States had no immediate plans for building such things. The Tonkas and Visols were no improvements on monoethylaniline, or on the aniline-furfuryl alcohol mixtures that had been developed in the U.S. And there was nothing new about nitric acid. The Americans thought they knew all about it—as had the Germans. Unwarranted euphoria and misplaced confidence are international phenomena.

They had no doubt that missiles, guided and ballistic, were to be the artillery of the future. The question—or one of many—was the identity of the optimum propellant combination for a given, or projected, missile. And so everybody even remotely connected with the business made his own survey of every conceivable fuel and oxidizer, and tried to decide which ones to choose. Lemmon, of JPL, presented the results of such a comprehensive survey to the Navy in the spring of 1945, and a half a dozen more, by North American Aviation, Reaction Motors, the Rand Corporation, M. W. Kellogg Co., and others, appeared in the next few years. Each survey listed the characteristics of every propellant, or prospective propellant, that the compiler could think of, and presented the results of dozens of tedious performance calculations. To the surprise of nobody with any chemical sophistication at all, everybody came to just about the same conclusions.

There were two sets of these. The first related to long-range ballistic missiles, or to rockets designed to orbit an artificial satellite. (As early as 1946 both the Air Force and the Navy were making serious studies of the problem of orbiting an artificial Earth satellite.) In these applications, cryogenics (substances that cannot be liquefied except at very low temperatures) could be used. And here everybody agreed that:

1 The optimum oxidizer is liquid oxygen. ("Fluorine might be good, but its density is too low, and it's a holy terror to handle.")

2 As far as performance is concerned, liquid hydrogen is tops as a fuel. (But it was extremely hard to handle, and to come by, and its density is so low that the necessary tankage would be immense.) Below hydrogen it didn't much matter. Alcohol, gasoline, kerosene—they'd all work pretty well, and could be lived with. ("But may be somebody could do something with things like diborane and pentaborane?" Their performances, as calculated, looked awfully impressive. "Sure, they were rare and expensive and poisonous to boot, but—?")

The second set of conclusions—or the lack of them—concerned things like JATO's and short range tactical missiles, which had to use storable propellants. Here the conclusions were less definite.

1 The available oxidizers were nitric acid, hydrogen peroxide (as soon as it could be got into production in the United States) and nitrogen tetroxide. (But N_2O_4 and 90 percent peroxide both froze at $-11°C$, and if you want to fight a war in, say, Siberia in February, or in the stratosphere—?) It looked as though nitric acid, in one of its variants, was the most likely candidate. ("Of course, if the freezing points of the other two could be reduced somehow—? And what about weirdies like ClF_3—?")

2 The conclusions were much less clean-cut when storable fuels were considered. With few exceptions, none of the possible fuels had a

performance much better than any of the others. Decisions would have to be made based on their secondary characteristics: availability, hypergolicity, smoothness of combustion, toxicity, and so on. The one important exception was hydrazine. (Not the hydrazine hydrate the Germans had been using, but anhydrous N_2H_4. Dave Horvitz, at RMI, fired the hydrate with oxygen in 1950, but I am not aware of any other experiments, in this country at least, in which it was involved. Almost all the hydrazine hydrate looted from Germany was converted to the anhydrous base before being distributed for testing. One method of conversion was to reflux the hydrate over barium oxide, and then to distil over the anhydrous hydrazine under reduced pressure.) Hydrazine was hypergolic with the prospective oxidizers, it had a high density for a fuel (1.004) and its performance was definitely better than those of the other prospective fuels. But—its freezing point was 1.5°C higher than that of water! And it cost almost twenty dollars a pound. So two things obviously had to be done—get the price of hydrazine down, and somehow, lower the freezing point. (And again, there was that haunting thought of pentaborane—?)

There was one subject on which everybody agreed. Nobody was going to put up with the aniline-RFNA combination for one moment longer than he had to. The acid was so corrosive to anything you wanted to make propellant tanks out of that it had to be loaded into the missile just before firing, which meant handling it in the field. And when poured it gives off dense clouds of highly poisonous NO_2, and the liquid itself produces dangerous and extremely painful burns when it touches the human hide. And . . . but nitric acid and the struggle to domesticate it deserve, and will get, a chapter all to themselves.

The aniline is almost as bad, but a bit more subtle in its actions. If a man is spashed generously with it, and it isn't removed immediately, he usually turns purple and then blue and is likely to die of cyanosis in a matter of minutes. So the combination was understandably unpopular, and the call went out for a new one that was, at least, not quite so poisonous and miserable to handle.

Kaplan and Borden at JPL suggested one at the beginning of 1946. This was WFNA and straight furfuryl alcohol. Furfuryl alcohol was about as harmless as any propellant was likely to be, and WFNA, while it was just as corrosive as RFNA, and was just as hard on the anatomy, at least didn't give off those clouds of NO_2. They fired the combination in a WAC Corporal motor, comparing it to the 20 percent furfuryl alcohol, 80 percent aniline mixture and RFNA, and found no measurable difference in performance between the two systems. (The WAC Corporal was conceived as a sounding rocket, the "Little Sister" to the 20,000 pound thrust "Corporal" then under development. It was the ancestor of the Aerobee.) And, as a bonus, they found that ignition was fast and smooth, and much more tolerant to water in the acid than was the Corporal combination.

At about the same time, RMI was making a similar set of tests. These were all run in a 220-pound thrust Lark motor, whose mixed-acid, monoethyl-aniline combination was the reference propellant system. They used three fuels—80 octane gasoline, furfuryl alcohol, and turpentine; and three types of nitric acid oxidizer—mixed acid, WFNA, and RFNA containing 15% N_2O_4.[*] They used a hypergolic starting slug on the gasoline firings, and rather surprisingly, got good results with all three acids. Furfuryl alcohol was no good with mixed acid. The combination was smoky and messy, and the reaction of the sulfuric acid of the MA with the alcohol produced a weird collection of tars, cokes, and resins, which quite clogged up the motor. But furfuryl alcohol was excellent with RFNA and WFNA, starting considerably smoother than did their reference propellants. And turpentine gave hard starts with RFNA and WFNA, but with MA started off like a fire hose. So that was one of the two combinations that they preferred. The other was furfuryl alcohol and WFNA (the RFNA performed a little better, but those NO_2 fumes!), although neat furfuryl alcohol freezes at $-31°C$—rather too high for comfort.

Many other fuels were tried during the late 40's and early 50's. At JPL mixtures of aniline with ethanol or with isopropanol were investigated and burned with RFNA. Ammonia was fired there (with RFNA) as early as 1949, and the next year Cole and Foster fired it with N_2O_4. The M. W. Kellogg Co. burned it with WFNA, and by 1951, R. J. Thompson of that company was beating the drum for this combination as the workhorse propellant for all occasions. Reaction Motors experimented with mixtures of ammonia and methylamine (to reduce the vapor pressure of the ammonia) and showed that the addition of 1.5 percent of dekaborane made ammonia hypergolic with WFNA, while the Bendix Corp., in 1953, showed that the same end could be achieved by flowing the ammonia over lithium wire just upstream of the injector.

JPL fired various oddities with RFNA, such as furfural and two methyl-ated and partially reduced pyridines, tetrapyre and pentaprim. The object of these tests is not readily apparent, nor is the reason why RMI bothered to fire cyclooctatetraene with WFNA. The fuel is not only expensive and hard to get, but it has a very high freezing point and has nothing in particular to recommend it. And the reason that the Naval Air Rocket Test Station went to the trouble of burning ethylene oxide with WFNA is equally baffling. The Edisonian approach has much to recommend it, but can be run into the ground. One of the oddest combinations to be investigated was tried by RMI, who burned d-limonene with WFNA. d-limonene is a terpene which can be

---

[*] Interestingly enough, the first stage of Diamant, which put the first French satellite into orbit, burns turpentine and RFNA.

extracted from the skins of citrus fruits, and all during the runs the test area was blanketed with a delightful odor of lemon oil. The contrast with the odors of most other rocket propellants makes the event worth recording.

It had long since become obvious to everybody concerned that firing a combination in a rocket motor is not the ideal way to find out whether or not it is hypergolic—and, if it is, how fast it ignites. By the nature of research more tests are going to fail than are going to succeed, and more combinations are going to ignite slowly than are going to light off in a hurry. And when the result of each delayed ignition is a demolished motor, a screening program can become a bit tedious and more than a bit expensive. So the initial screening moved from the test stand into the laboratory, as various agencies built themselves ignition delay apparatus of one sort or another. Most of these devices were intended not only to determine whether or not a combination was hypergolic, but also to measure the ignition delay if it was. In construction they varied wildly, the designs being limited only by the imagination of the investigator. The simplest tester consisted of an eyedropper, a small beaker, and a finely calibrated eyeball—and the most complicated was practically a small rocket motor setup. And there was everything in between. One of the fancier rigs was conceived by my immediate boss, Paul Terlizzi, at NARTS. He wanted to take high-speed Schlieren (shadow) movies of the ignition process. (What information he thought they would provide escaped me at the time, and still does.)* There was a small ignition chamber, with high-speed valves and injectors for the propellants under investigation. Viewing ports, a high-speed Fastex camera, and about forty pounds of lenses, prisms, and what not, most of them salvaged from German submarine periscopes, completed the setup. Dr. Milton Scheer (Uncle Milty) labored over the thing for weeks, getting all the optics lined up and focused.

Came the day of the first trial. The propellants were hydrazine and WFNA. We were all gathered around waiting for the balloon to go up, when Uncle Milty warned, "Hold it—the acid valve is leaking!"

"Go ahead—fire anyway!" Paul ordered.

I looked around and signaled to my own gang, and we started backing gently away, like so many cats with wet feet. Howard Streim opened his mouth to protest, but as he said later, "I saw that dog-eating grin on Doc's face and shut it again," and somebody pushed the button. There was a little flicker of yellow flame, and then a brilliant blue-white flash and an ear-splitting crack. The lid to the chamber went through the ceiling (we found it in the attic some

---

* An incurable inventor of acronyms, he called it "STIDA," for Schlieren Type Ignition Delay Apparatus. MHF-3, introduced by Reaction Motors a few years ago, is 86% monomethyl hydrazine and 14% hydrazine. "And there is nothing new under the sun."

weeks later), the viewports vanished, and some forty pounds of high-grade optical glass was reduced to a fine powder before I could blink.

I clasped both hands over my mouth and staggered out of the lab, to collapse on the lawn and laugh myself sick, and Paul stalked out in a huff. When I tottered weakly back into the lab some hours later I found that my gang had sawed out, carried away, and carefully lost, some four feet from the middle of the table on which the gadget had rested, so that Paul's STIDA could never, never, never be reassembled, in *our* lab.

Other agencies had their troubles with ignition delay apparatus, although their experiences weren't often as spectacular as ours, but they eventually started cranking out results. Not too surprisingly, no two laboratories got the same numbers, and from 1945 until 1955 one would be hard put to find a period when there wasn't a cooperative ignition delay program going on, as the various laboratories tried to reconcile their results. One of the difficulties was that the different testers varied widely in the speed and the efficiency with which they mixed the two reactants. And another lay in the fact that different criteria for ignition were used by various experimenters. One might take the first appearance of flame (as shown by a photo cell or an ionization gage or a high-speed camera) as the moment of ignition, while another, with a micro-motor setup, might take the moment at which his motor arrived at full thrust or the design chamber pressure.

But although the various investigators didn't often come up with the same numbers, they generally rated propellant combinations in the same order. While they seldom agreed on the number of milliseconds it took combination A to light off, they were generally in complete agreement that it was a Hell of a lot faster than combination B.

Which was enough for many purposes. After all, everybody knew that WFNA and furfuryl alcohol were fast enough to live with, and obviously, if something shows up on the tester as faster than that combination it's probably worth trying in a motor.

Many laboratories worked in the field but Don Griffin at JPL and Lou Rapp at RMI were early comers in ignition delay work. The former organization, as was natural since Corporal was their baby, did a lot of work on the aniline-furfuryl-alcohol mixture, and in 1948 determined that the mixture with the minimum ignition delay consisted of 60 percent of the alcohol and 40 of aniline. This was close to the 49FA, 51 aniline eutectic (melting point −43°C) and the Corporal fuel (the missile was still under development) was changed from the 20 percent FA mixture to a 50–50 one.

Otherwise, they confirmed the hypergolic reaction of furan compounds and of aromatic amines with nitric acid, and demonstrated the beneficial effect of N_2O_4 in the latter case. And they showed that amines, particularly tertiary amines, and unsaturated compounds were generally hypergolic, while aliphatic alcohols and saturates generally were not. Most of their work was done with nitric acid, but a good deal, from 1948 on, was done with N_2O_4, whose hypergolic nature generally resembled that of that acid.

Reaction Motors investigated the hypergolicity of similar compounds, as well as such things as the furans, vinyl and allyl amines, and polyacetylenics, such as di-propargyl, with the skeleton structure (without the hydrogens) $C \equiv C — C — C — C \equiv C$. And they found that many silanes were hypergolic with acid. The University of Texas, in 1948, also worked with these, and showed that 30 percent of tetra-allyl silane would make gasoline hypergolic. The University of Texas also investigated the zinc alkyls, as Sänger had done sixteen years earlier.

Standard Oil of California was the first of the oil companies to get into rocket propellant research in a big way, when Mike Pino, at the company's research arm, California Research, started measuring ignition delays in the fall of 1948.

At first his work resembled that of the other workers, as he demonstrated fast ignition with dienes, acetylenics, and allyl amines. (Some years later, in 1954, Lou Rapp at RMI assembled the results of all the early ignition delay work, and attempted to make some generalizations. His major conclusion was that the ignition of a hydrocarbon or an alcohol involved the reaction of the acid with a double or triple bond, and that if none existed it had to be created before the ignition could take place. Later, in speaking of nitric acid, the plausibility of this postulate will be examined.)

But then Pino, in 1949, made a discovery that can fairly be described as revolting. He discovered that butyl mercaptan was very rapidly hypergolic with mixed acid. This naturally delighted Standard of California, whose crudes contained large quantities of mercaptans and sulfides which had to be removed in order to make their gasoline socially acceptable. So they had drums and drums of mixed butyl mercaptans, and no use for it. If they could only sell it for rocket fuel life would indeed be beautiful.

Well, it had two virtues, or maybe three. It was hypergolic with mixed acid, and it had a rather high density for a fuel. And it wasn't corrosive. But its performance was below that of a straight hydrocarbon, and its odor—! Well, its odor was something to consider. Intense, pervasive and penetrating, and resembling the stink of an enraged skunk, but surpassing, by far, the best efforts of the most vigorous specimen of *Mephitis mephitis*. It also clings to the clothes and the skin. But rocketeers are a hardy breed, and the stuff was duly and successfully fired, although it is rumored that certain rocket mechanics were excluded from their car pools and had to run behind. Ten years after it was fired at the Naval Air Rocket Test Station—NARTS—the odor was still noticeable around the test areas. (And at NARTS, with more zeal than judgment, I actually developed an analysis for it!)

California Research had an extremely posh laboratory at Richmond, on San Francisco Bay, and that was where Pino started his investigations. But when he started working on the mercaptans, he and his accomplices were exiled to a wooden shack out in the boondocks at least two hundred yards from the main building. Undeterred and unrepentant, he continued his noisome endeavors, but it is very much worth noting that their emphasis had changed. His next candidates were not petroleum by-products, nor were they chemicals which were commercially available. They were synthesized by his own crew, specifically for fuels. Here, at the very beginning of the 50's, the chemists started taking over from the engineers, synthesizing new propellants (which were frequently entirely new compounds) to order, instead of being content with items off the shelf.

Anyhow, he came up with the ethyl mercaptal of acetaldehyde and the ethyl mercaptol of acetone, with the skeleton structures:

$$C-C-S-\underset{\underset{C}{|}}{C}-S-C-C \quad \text{and} \quad C-C-S-\overset{\overset{C}{|}}{\underset{\underset{C}{|}}{C}}-S-C-C$$

respectively. The odor of these was not so much skunk-like as garlicky, the epitome and concentrate of all the back doors of all the bad Greek restaurants in all the world. And finally he surpassed himself with something that had a dimethylamino group attached to a mercaptan sulfur, and whose odor can't, with all the resources of the English language, even be described. It also drew flies. This was too much, even for Pino and his unregenerate crew, and they banished it to a hole in the ground another two hundred yards farther out into the tule marshes. Some months later, in the dead of night, they surreptitiously consigned it to the bottom of San Francisco Bay.

To understand the entry of the next group of workers into the propellant field, it's necessary to go back a bit and pick up another thread. From the beginning, the services had disliked the fuels that the researchers had offered them, not only because of their inherent disadvantages, but above all because they weren't gasoline. They already had gasoline and used huge quantities of it—and why should they have to bother with something else? But, as we have seen, gasoline is not a good fuel to burn with nitric acid, and the services had to accept the fact. Which they did, grudgingly. But all through the late 40's and early 50's the Navy and the Air Force were busily changing over from piston airplane engines to turbojets. And they started buying jet fuel instead of gasoline, and the whole thing started all over again. They demanded of the people designing their missiles that said missiles be fueled with jet fuel.

Now, what is jet fuel? That depends. A turbojet has a remarkably undiscriminating appetite, and will run, or can be made to run, on just about anything that will burn and can be made to flow, from coal dust to hydrogen. But the services decided, in setting up the specifications for the jet fuel that they were willing to buy, that the most important considerations should be availability and ease of handling. So since petroleum was the most readily available source of thermal energy in the country, and since they had been handling petroleum products for years, and knew all about it, the services decided that jet fuel should be a petroleum derivative—a kerosene.

The first fuel that they specified was JP-1, a rather narrow cut, high paraffinic kerosene. The oil companies pointed out that not many refineries in the country could produce such a product with their available equipment and crudes, and that the supply might thus be somewhat limited. So the next specification, for JP-3 (JP-2 was an experimental fuel that never got anywhere), was remarkably liberal, with a wide cut (range of distillation temperatures) and with such permissive limits on olefins and aromatics that any refinery above the level of a Kentucky moonshiner's pot still could convert at least half of any crude to jet fuel. This time they went too far, allowing such a large fraction of low boiling constituents that a jet plane at high altitude boiled off a good part of its fuel. So the cut was narrowed to avoid this difficulty, but the permitted fractions of aromatics and olefins (25 and 5 percent respectively) were not reduced. The result was JP-4, with just about the most permissive specifications to appear since the days of Coal Oil Johnny Rockefeller the First. It is NATO standard, and the usual fuel for everything from a Boeing 707 to an F-111. (JP-5 and 6 have arrived since, but haven't replaced JP-4. And RP-1 is another story, which will be told later.)

But trying to burn JP-3 or JP-4 in a rocket motor with nitric acid was a harrowing experience. In the first place, the specifications being what they were, no two barrels of it were alike. (A jet engine doesn't care about the shape of the molecules it burns as long as they give up the right number of BTU's per pound, but a nitric acid rocket is fussier.) It wasn't hypergolic with acid, but reacted with it to produce all sorts of tars, goos, weird colored compounds of cryptic composition—and troubles. And if you got it going—using a hypergolic slug, say—sometimes everything went well, but usually not. It was acid-gasoline all over again—a coughing, choking, screaming motor, that usually managed to reduce itself to fragments, and the engineers to frustrated blasphemy. Everything was tried to make the stuff burn smoothly, from catalysts in the acid down—or up—to voodoo. The farthest-out expedient that I heard of was tried at Bell Aeronautic. Somebody had the bright idea that the sonic vibrations of a rocket motor might promote combustion. So he made a tape recording of the sound of a running motor and played it back at the interacting propellants in the hope that they might be shaken—or shamed—into smooth combustion. (Why not? He'd tried everything else!) But alas, this didn't work either. Obviously JP was a lost cause as far as the rocket business was concerned.

It was with this background that the Navy's program on "Rocket Fuels Derivable from Petroleum" came into being in the spring of 1951, although it wasn't called that officially until the next year. If you couldn't make JP work, maybe you could derive something else (cheaply, for choice) from petroleum that would. Or, one hoped, that could be mixed with JP and make the latter burn smoothly over a reasonable mixture-ratio range.

The title of the program was deceptive. "Derivable" is an elastic term, and it is to be doubted that the higher-ups of the Bureau of Aeronautics realized what they had authorized. But the lower-level chemist types in the Rocket Branch were perfectly aware of the fact that a good chemist, given a little time and money, can derive just about anything organic, up to RNA, from petroleum if he wants to. The contractors were being told, in effect, "Go ahead, Mack—see what you can come up with. And if it's any good, we'll find a way to make it from petroleum—somehow!"

The contractors now joining their endeavors to those of California Research were the Shell Development Co., Standard Oil of Indiana, Phillips Petroleum, and the Chemical Engineering Department of New York University (NYU). And for the next two or three years there was a continuous ignition delay project going on. Each laboratory, as it came up with a new hypergolic additive, would ship samples to all the others, who would mix it with standard nonhypergolic fuels and then measure the ignition delay of the mixtures. The standard nonhypergols were generally toluene and n-heptane, although NYU, presumably to assert its academic independence, used benzene and n-hexane. (JP wasn't much use as a reference fuel, since no two lots of it were alike.)

As for the fuels and/or additives that they synthesized, Shell and NYU concentrated on acetylenic compounds, and Phillips put their major effort into amines. As for Standard of Indiana, that organization went off on a wild tangent. Apparently jealous of their sister company of California and determined to do them one better, they went beyond mere sulfur compounds, and came down hard on phosphorous derivatives. They investigated assorted substituted phosphines, from the timethyl phosphine, through butyl and octyl phosphines, on to monochloro (dimethylamino) phosphine, and then they settled happily on the alkyl trithiophosphites, with the general formula $(RS)_3P$, where R could be methyl, ethyl, or whatever. The one they gave the greatest play was "mixed alkyl trithiophosphites," which was a mixture of, mainly, the ethyl and methyl compounds. Its virtues were those of the mercaptans—hypergolicity and good density and no corrosion problems—but its vices were also those of the mercaptans—exaggerated. The performance was below that of the mercaptans, and the odor, while not as strong as those of the Pino's creations, was utterly and indescribably vile. Furthermore, their structures had an unnerving resemblance to those of the G agents, or "nerve gases" or of some of the insecticides which so alarmed Rachel Carlson. This disquietude was justified. When some of the alkylthiophosphites were fired at NARTS, they put two rocket mechanics in the hospital, whereupon they were summarily and violently thrown off the station. Standard of Indiana plugged them hard, and there was even a conference devoted to them in March of 1953, but somehow they, like the mercaptans, never roused the enthusiasm of the prospective users. Neither type of propellant, now, is anything but a noisome memory.

The rationale behind the acetylenic work was clear enough. It had been shown (by Lou Rapp and Mike Pino, among others) that double and triple bonds aided hypergolic ignition, and it was reasonable to assume that they might promote smooth combustion, if only by furnishing the fuel molecule with a weak point where the oxidation might start. Furthermore, the parent molecule of the family, acetylene itself, had always been regarded hopefully by the workers in the field. The extra energy conferred upon it by the triple bond should lead to good performance, although the low percentage of hydrogen in the molecule might work against it. (See the chapter on performance.) But pure liquid acetylene was just too dangerous to live with—having a lamentable tendency to detonate without warning and for no apparent reason. Perhaps some of its derivatives might be less temperamental. And these was another reason for looking at the acetylenics.

A good many people, in the early 50's, were considering some unusual, not to say bizarre, propulsion cycles. Among these was the ram rocket. This is a rocket, generally a monopropellant rocket, inside of and surrounded by a ramjet. A ramjet will not function except at high speed relative to the atmosphere, and hence has to be boosted into operation by a rocket or some other means. If the enclosed rocket of the ram rocket could get the device up to operating velocity, and if the rocket exhaust gases were combustible and could act as the fuel for the ramjet—well, then you could build a cruising missile that didn't need a booster and with a lower specific fuel consumption than a straight rocket. Say that you burned propyne, or methyl acetylene, in a monopropellant rocket, and that the exhaust products were largely methane and finely divided elementary carbon. Then the carbon and the methane could be burned with air in the ramjet, going to water and carbon dioxide, and you would be making the best of both worlds. (Ethylene oxide, C_2H_4O, whose major decomposition products are methane and carbon monoxide, was considered for the same sort of cycle.) So the acetylenics looked good for the ram rocket.

And finally, the acetylenics are rather easy to produce from petroleum feedstock, by cracking and partial oxidation. The approaches of NYU and of Shell to the acetylenic problem were completely dissimilar. NYU tried dozens of compounds of the family, while Shell concentrated on just two, and then went hunting for additives which would make them into useful fuels. One of the two was 1,6-heptadiyne, with the skeletal structure $C \equiv C - C - C - C - C \equiv C$. And the other was 2-methyl-1-buten-3-yne, otherwise known as "isopropenyl acetylene" or "methyl vinyl acetylene," whose skeleton is

$$C=C-C\equiv C$$

(with a C branch off the second carbon). One source of confusion in the history of the acetylenics is the multiplicity of systems by which they were named!

The first additives that they investigated thoroughly were methyl derivatives of phosphorous triamide, $P(NH_2)_3$, with methyl groups substituted for from three to six of the hydrogens. They worked, but so much of the additive was needed for proper ignition that it became a major component of the mixture, and even then explosive ignition was common.

Then they tried the derivative of 1,3,2-dioxaphospholane,

and finally settled on 2-dimethylamino-4-methyl-1,3,2-dioxaphospholane, which was usually, and mercifully, known as "Reference Fuel 208." Again, it wasn't a success as an additive, but taken neat, it was one of the fastest hypergols ever seen. It wasn't particularly toxic, and might have made a fairly good workhorse fuel, but before much work had been done on it, events made it obsolete. It's all but forgotten now.

Between 1951 and 1955 Happell and Marsel at NYU prepared and characterized some fifty acetylenics: hydrocarbons, alcohols, ethers, amines, and nitriles. They varied in complexity from propyne, or methyl acetylene, $C-C\equiv C$ to such things as dimethyldivinyldiacetylen

$$C=C-C\equiv C-C\equiv C-C=C$$

(with C branches) with no less than four multiple bonds. The climax of unsaturation came with butyne di-nitrile, or dicyanoacetylene, $N\equiv C-C\equiv C-C\equiv N$ which had no hydrogen atoms at all, but rejoiced in the possession of three triple bonds. This was useless as a propellant—it was unstable, for one thing, and its freezing point was too high—but it has one claim to fame. Burning it with ozone in a laboratory experiment, Professor Grosse of Temple University (who always liked living dangerously) attained a steady state temperature of some 6000 K, equal to that of the surface of the sun.

Many, if not most, of the acetylenics had poor storage properties, and tended to change to tars or gels on standing. They also tended to form explosive peroxides on exposure to the atmosphere. Many of them were shock sensitive, and would decompose explosively with little or no provocation. Something like divinyldiacetylene can fairly be described as an accident looking for a place to happen. While some of them were fired successfully in a rocket (RMI burned propyne, methylvinylacetylene methyldivinylacetylene, and dimethyl-divinylacetylene, all with oxygen) they turned out not to be suitable fuels for nitric acid. They usually detonated on contact with the oxidizer, as several possessors of piles of junk that had originally been ignition delay equipment could testify, and did.

But some of them showed promise as monopropellants and as additives, and the Air Reduction Co., which had entered the field around the middle of 1953, had propyne, methylvinylacetylene, and dimethyldivinylacetylene in commercial production by 1955.

Some of them were excellent additives for JP-4. By August, 1953, RMI had shown that as little as 10 percent of methylvinylacetylene in JP-4 led to smooth combustion with RFNA over a wide range of mixture ratios, and greatly improved ignition. If a hypergolic slug was used, transition to the working fuel was smooth and without incident, and, for that matter, ignition could easily be achieved with a powder squib, and without a starting slug at all. Several of the others had the same effect, but by the time that this was determined the acetylenics had been overtaken by history, and had been developed only to be abandoned.

Homer Fox and Howard Bost ran the amine program at Phillips Petroleum. The relationship of amines to petroleum is exiguous at best, but they had been used as fuels for some time (triethylamine had been used in the Tonkas) and looked good, although they had never been examined systematically for propellant use. This Phillips proceeded to do, and investigated amines in infinite variety. Primary, secondary, and tertiary amines. Saturated and unsaturated amines, allyl and propargyl amines. Monoamines, diamines, even triamines and tetramines. They must have synthesized and characterized at least forty aliphatic amines, including a few with other functional groups—OH groups and ether linkages.

They concentrated on the tertiary polyamines. This was logical enough. They knew that tertiary amines were generally hypergolic with nitric acid, and it was reasonable to think that a di- or tri-tertiary amine might be more so. (Their guess turned out to be right, but one is reminded of E. T. Bell's remark that the great vice of the Greeks was not sodomy but extrapolation.) The compounds they investigated ranged from 1,2 bis (dimethylamino) ethane, up to such curiosities as 1,2,3, tris (dimethylamino) propane and tetrakis (dimeth-ylaminomethyl) methane, which can be visualized as a neopentane molecule with a dimethylamine group on each corner. Incidentally, it turned out to have an unacceptably high freezing point, which, considering the symmetry of the molecule, might have been expected. One is led to suspect that some of the fancier amines were synthesized, not because there was any reason to believe that they would be an improvement on the ones they already had, but to demonstrate the virtuosity of the bench man, who wanted to prove that he could do it.

The tertiary diamines were the ones that really got a workout. Just about every possible structural change, and its consequences, were investigated. Thus they investigated the consequences of varying the terminal groups, as in the series:

> 1,2 bis (dimethyl, or ethyl, or allylamino) ethane.

Or, of varying the length of the central hydrocarbon chain, as in:

$$
\left.\begin{array}{l} 1,1 \\ 1,2 \\ 1,3 \\ 1,4 \\ 1,6 \end{array}\right\} \text{-bis (dimethylamino)} \left\{\begin{array}{l} \text{methane} \\ \text{ethane} \\ \text{propane} \\ \text{butane} \\ \text{hexane} \end{array}\right.
$$

They moved the amino groups around, as in:

$$
\left.\begin{array}{l} 1,2 \\ 1,3 \end{array}\right\} \text{bis (dimethylamino)-propane and}
$$

$$
\left.\begin{array}{l} 1,2 \\ 1,3 \\ 1,4 \end{array}\right\} \text{bis (dimethylamino) butane.}
$$

They examined the effect of unsaturation, in series like

$$
1,4 \text{ bis (dimethylanimo)} \left\{\begin{array}{l} \text{butane} \\ 2 \text{ butene} \\ 2 \text{ butyne} \end{array}\right.
$$

And they tried every conceivable permutation and combination of these changes, as well as adding OH groups or ether linkages.

As might have been expected, introducing an hydroxyl group produced a compound which was excessively viscous at low temperatures. (Triethanolamine, which had been considered as a fuel, is an extreme example of this effect, and therefore was never used.) The allyl-terminated amines were also rather viscous, and were subject to atmospheric oxidation. Otherwise, as might have been expected, they were all very much alike, the complicated ones being in no way superior to the simple compounds, as might also have been expected.

None of them was any good as a jet fuel additive. They neither improved combustion nor, except in overwhelming proportions, made the jet fuel hypergolic. However, they looked promising as straight fuels, and Phillips shipped samples of four of them to the Wright Air Development Center to be test fired. They were all of the bis (dimethylamino) type, the 1,2 ethane, the 1,2, and 1,3 propane, and the 1,3-1 butene.

At WADC, in 1956, Jack Gordon checked out their properties and logistics, and fired them with RFNA. They were good fuels. Ignition was hypergolic and fast, combustion was good and performance was respectable, and the saturated ones, at least, were quite stable to heat and suitable for regenerative cooling.

And they, too, were obsolete at birth.

For all this work had been done, as it were, with the left hand. Hydrazine was the name of the big game. That was the fuel that everybody wanted to use. High performance, good density, hypergolic with the storable oxidizers—it had everything. Almost,

Its price was high, but the nature of the chemical industry being what it was, and is, one could be confident that it would come down to a reasonable figure when anybody wanted it in quantity. It was somewhat sensitive to catalytic decomposition, but if you used the right materials to make your tanks of, and were reasonably careful about cleanliness, that was no real problem. But that freezing point—1.5°C—was just too high for anything that was going to be used in a tactical missile. The services were awfully coy about setting definite limits on the freezing point of propellants that they would accept—one had the feeling that they would demand the impossible and then settle for what they could get—but they finally decided that −65°F, or −54°C, would be acceptable for most purposes. (Although the Navy, during one whimsical period, demanded a freezing point no higher than −100°F. How they would fight a war at that temperature they didn't specify. One is tempted to believe that they were carried away by the magnificent evenness of the number.)

So everybody was trying to bring the freezing point of hydrazine down to −54°. And without adversely affecting its other—and good—properties. Which turned out to be impossible. This could have been predicted, but at that time we were all hoping for miracles.

From first to last, at least eight agencies were involved in the effort, Aerojet, JPL, the Metalectro Co., NARTS, Naval Ordnance Test Station (NOTS), North American Aviation, Reaction Motors, and Syracuse University.

The first freezing point depressant to be tried—although involuntarily—was water. Hydrazine hydrate, which is 36 percent water, has a freezing point of −51.7°, and a mixture containing 42 percent water freezes at −54°. (V. I. Semishin, in Russia, had determined part of the hydrazine-water phase diagram in 1938, and Mohr and Audrieth, in this country, in 1949, and Hill and Summer, in England, in 1951 completed the job.) But water was an extremely bad additive for a fuel. It contributed nothing to the energetics of the system, and the mass of the water, just going along for the ride, seriously degraded the performance.

Ammonia wasn't quite as bad. F. Fredericks, in 1913 and in 1923, had reported on the hydrazine-ammonia phase diagram, which was also investigated by D. D. Thomas, at JPL, in 1948. Ammonia, unlike water, was a fuel, but it is a very stable compound and its heat of combustion is not what might be desired. And it took something like 61 percent of ammonia in the hydrazine to reduce its freezing point to $-54°$! This not only reduced the performance sharply, but decreased the density of the fuel, and, on top of that increased its vapor pressure so much that it boiled at about $-25°$ instead of at the $+113.5°$ boiling point of pure hydrazine. Dave Horvitz of RMI investigated ternary mixtures of hydrazine, water, and ammonia in 1950, but couldn't find any mixture that possessed both an acceptable freezing point and a large fraction of hydrazine. Water and ammonia were not the answers.

Another additive investigated by RMI (in 1947) was methanol. A mixture containing 44 percent of the hydrazine and 56 of the alcohol freezes at—$54°$, and its other physical properties are acceptable, but it yields a performance considerably below that of the neat hydrazine. Some years later, under circumstances which will be described, interest in the mixture was revived.

Don Armstrong, of Aerojet, came up with something in the summer of 1948 that for a time looked extremely promising. He found that the addition of 13 percent of lithium borohydride to hydrazine produced a mixture whose (eutectic) freezing point was $-49°$. Not the magic $-54°$, but still something. The density was reduced somewhat, from 1.004 to about 0.93, but as the borohydride itself is such an energetic compound there was no reason to expect any appreciable degradation of the performance. But, alas, his triumph was illusory. After some time had elapsed the mixture was found to be inherently unstable, and slowly and inexorably to decompose, with a steady evolution of hydrogen. They gave the whole idea up around 1952, but RMI was looking at it as late as 1958, and only around 1966 or 1967 somebody else suggested using $LiBH_4$ as a freezing point depressant for hydrazine! This may indicate something beyond a profound and depressing ignorance of the history of one's own technology, but I'm not exactly sure what.

At about the same time T. L. Thompson, of North American, came up with another freezing point additive, whose major drawback, although its thermal stability was poor, was that it scared everybody to death. He found that 15 percent of hydrocyanic acid, HCN, would reduce the freezing point of hydrazine to $-54°$. But the mere thought of the HCN so alarmed everybody (although much more toxic compounds had been and would be investigated, and without any particular notice being taken of it) that the mixture was never accepted.

At about this time (1949–50) the LAR missile was being developed at NOTS, and E. D. Campbell and his associates came up with a low freezing fuel for it—a mixture of 67 percent hydrazine and 33 ammonium thiocyanate, with a freezing point of −54°. This could be lived with, although the performance was somewhat degraded and the vapor pressure was inconveniently high.

Early in 1951, Dave Horvitz at the Metallectro Co. (where he had moved from RMI) investigated hydrazine-aniline mixtures, and found that the eutectic composition, with a freezing point of −36°, contained only 17 percent hydrazine. He then started adding methylamine to the mixture, to reduce the viscosity as well as the freezing point, and finally came up with a hydrazine-aniline-methylamine mixture (regrettably called "HAM Juice"), which froze at −50°, but contained only 9.1 percent of hydrazine, with 19.3 of methylamine and 71.6 of aniline. This was investigated rather thoroughly, and was test fired, but it wasn't the answer that people were looking for. (But the Army, in 1953 added 5 percent of hydrazine to their aniline-furfuryl alcohol Corporal fuel, and three years later raised the percentage to seven.)

One of the most thoroughly investigated additives was hydrazine nitrate. The ammonia analogue of the mixture—ammonium nitrate in ammonia, Diver's solution—had been around for years, so the idea was obvious enough, and apparently several people thought of it independently at about the same time. Dwiggins at the Naval Ordnance Laboratory (NOL) and my group at NARTS investigated the system in 1951, and by the end of 1953 J. M. Corcoran and his colleagues at NOTS had worked out the whole hydrazine–hydrazine nitrate–water system. A mixture containing 55 percent hydrazine and 45 hydrazine nitrate froze below −40°, and the magic −54° could be attained with one containing 54 percent hydrazine, 33 of the nitrate, and 13 water. This was not bad, but there was, as usual, a catch or two. The mixtures were quite viscous at low temperatures, and had a tendency to froth, which could lead to trouble if a pumped feed system were used. And, particularly, most of the really useful mixtures, with low percentages of water, could be detonated with alarming ease. (And the dry hydrazine nitrate, if mistreated, could produce a very plausible simulation of a turret fire. The NARTS group found that out!) But some of the mixtures could be used as monopropellants, and as such, were studied extensively for some years, and some of them were tried as liquid gun propellants.

The NARTS group, not content with mere nitrates, tried hydrazine perchlorate as a depressant in 1951, and found that a mixture containing 49 percent hydrazine, 41.5 of the perchlorate, and 8.5 water was still liquid at −54°. But it was even more likely to detonate than were the nitrate mixtures (while attempting to investigate its thermal stability we blew a hole in the ceiling of the laboratory), and I discovered, nearly blowing my head off in the process, that it is not advisable to attempt to dehydrate the hydrazinium perchlorate hemi-hydrate (the form in which it crystallizes) to the anhydrous salt. So, although the perchlorate mixture was more energetic than the nitrate mixtures, its use was outside of the range of practical politics. Nevertheless, Walker, at Syracuse University, tried sodium perchlorate monohydrate a year or so later, and found that a 50-percent mixture with hydrazine froze at approximately $-46°$. Somehow, he managed to do it without killing himself.

Many other freezing point depressants were tried by various groups, with little or no success, and it was rapidly becoming obvious that the additive approach wasn't going to get anywhere. You either ruined your performance or were likely to blow your head off. Something new had to be added to people's thinking.

It was a Navy program that led to the breakout. At the beginning of 1951 the Rocket Branch of the Bureau of Aeronautics granted contracts to Metallectro and to Aerojet to synthesize certain hydrazine derivatives, and to determine their suitability as rocket propellants.

The three derivatives were monomethylhydrazine, symmetrical dimethyl hydrazine, and unsymmetrical dimethyl hydrazine. The hope was that a very slight alteration to the structure—and you can hardly alter it less than by adding a methyl group—might give it a reasonable freezing point without changing its energetics enough to matter.

At NARTS, I had the same idea, and managing to lay my hands on a pound of monomethylhydrazine—it cost $50.00—I investigated its mixtures with hydrazine, and before the end of the year recommended the eutectic, which contained 12 percent hydrazine and froze at $-61°$, as *the* fuel to concentrate on. The performance with HNO_3 was about 98 percent of that of straight hydrazine, the density was not too bad (0.89) the freezing point was beautiful, the viscosity was nothing to worry about, and storage and handling didn't seem to involve any particular problems, although the methyl hydrazine appeared to be a bit more sensitive to catalytic decomposition than the parent compound.

It didn't take Metallectro and Aerojet very long to discover that they were on to something good. Symmetrical dimethyl hydrazine turned out to be a dog (it's freezing point was only $-8.9°$), but monomethylhydrazine (to be referred to from now on as MMH) melted at $-52.4°$, and unsymmetrical dimethyl hydrazine (UDMH) melted at $-57.2°$. And Dave Horvitz at Metallectro found that the 60–40 UDMH-MMH eutectic mixture froze only at $-80°$, or $-112°F$, thus exceeding the Navy's mystic goal. What's more, its viscosity at their magic $-100°F$ was only 50 centipoises, so that it could really be used at that temperature. In the meantime, Aston and his colleagues at Pennsylvania State College had been determining the thermodynamic properties (heat of formation, heat capacity, heat of vaporization, etc.) of the substituted hydrazines, and by 1953 just about every useful piece of information about UDMH and MMH had been firmly nailed down.

They were both magnificent fuels—and the question that had to be decided was which one to concentrate on. A symposium on hydrazine and its derivatives and applications was held in February 1953, and the question was argued at length and with heat. MMH was a little denser than UDMH, and had a slightly higher performance. On the other hand, UDMH was less liable to catalytic decomposition, and had such good thermal stability that it could easily be used for regenerative cooling. Either one could be used as a combustion additive for JP-4, but UDMH was more soluble, and would tolerate a larger percentage of water in the fuel without separating. Both were hypergolic with nitric acid, the UDMH being the faster—after all, it was not only a hydrazine, but also a tertiary amine. And they both performed well as propellants, with performances superior to those of the tertiary diamines or of any of the phosphorous or sulfur compounds or of the old aniline type or furfuryl alcohol fuels. My MMH-hydrazine mixture was fired at NARTS early in 1954, UDMH at WADC at about the same time, MMH a little later and the UDMH-MMH eutectic at the same agency during 1955—all with red fuming nitric acid. And UDMH in JP-4 was so successful in smoothing out combustion that the fuel decided upon for the Nike Ajax missile was 17 percent UDMH in JP-4. The substituted hydrazine program was a resounding success. It had made all the other storable fuels completely obsolete.

The final decision to concentrate on UDMH was made on economic grounds. The two competitors for the first production contract for the substituted hydrazines were Metallectro and the Westvaco Chlor-Alkali division of Food Machinery and Chemical Co. (FMC). Metallectro proposed using a modification of the classic Raschig process for hydrazine, by reacting chloroamine with mono or dimethyl amine, according to which of the two hydrazines the customer wanted. And in their bid they proposed a carefully worked out sliding scale of prices, depending on the size of the order.

Westvaco took another approach. They proposed using another synthesis in which nitrous acid reacts with dimethyl amine to form nitrosodimethylamine, which can easily be reduced to the UDMH. The process cannot be used for MMH, and so Westvaco ignored the latter, and being prepared to take a loss on the initial orders (after all, the money involved was trivial from the point of view of a company the size of FMC), drastically underbid Metallectro. They got the order, and Metallectro dropped out of the picture for good. The first military specifications for UDMH were published in September 1955.

But that didn't inhibit Westvaco's advertising department. Intoxicated with success, and military specification or no, they tried to get away with a trade name, and called their stuff "DIMAZINE—the Westvaco brand of UDMH" and insisted that all of their people refer to it by that name. I pitied some of their chemists, visiting various agencies in the rocket business, dutifully and blushingly obeying orders, amidst the ribald hoots from their highly sophisticated audiences, who were as aware as they were themselves of the fact that Westvaco UDMH was absolutely indistinguishable from that made by Olin Mathieson or anyone else.

Some attempts were made to improve upon UDMH. Mike Pino at California Research had, as we have seen, worked with allyl amines, and in 1954 he carried this a bit further, and came up with the mono and the unsymmetrical diallyl hydrazines. These were interesting, but no particular improvement over UDMH, and were sensitive to oxidation and polymerization. And the people at Dow Chemical, a little later, produced monopropargyl hydrazine and unsymmetrical dipropargyl hydrazine. Again, no improvement, and both of them were horribly viscous at low temperatures. And McBride and his group, at NOTS, studying the oxidation chemistry of UDMH, in 1956 came upon tetramethyl tetrazene $(CH_3)_2N—N=N—N(CH_3)_2$. But its performance advantage over UDMH was trivial, and its freezing point was quite high.

So UDMH, for several years, was *the* fuel to be burned with nitric acid or N_2O_4. But, as designers have been trying to wring the last possible second of performance out of their motors, MMH has been growing in popularity. (It, too, has a Mil. Spec. now!) And, in applications which do not require a low freezing point, hydrazine itself is used, either straight or mixed with one of its derivatives. The fuel of the Titan II ICBM doesn't have to have a low freezing point, since Titan II lives in a steam-heated hole in the ground, but it *does* need the highest possible performance, and hydrazine was the first candidate for the job. But, as hydrazine has an unfortunate tendency to detonate if you try to use it as a regenerative coolant, the fuel finally chosen was a 50–50 mixture of hydrazine and UDMH, called "Aerozine 50" by Aerojet who came up with it first, and "50–50" by everybody else.

Today there are a bewildering lot of hydrazine-type fuels around, with names like MAF-3 (Mixed Amine Fuel-3) or MHF-5 (Mixed Hydrazine Fuel-5) or Hydyne, or Aerozine-50, or Hydrazoid N, or U-DETA or whatever. But whatever the name, the fuel is a mixture of two or more of the following: hydrazine, MMH, UDMH, diethylene triamine (DETA, added to increase the density), acetonitrile (added to reduce the viscosity of mixtures containing DETA) and hydrazine nitrate. And, for one special application (a vernier motor on Surveyor) enough water was added to MMH to form the monohydrate, whose cooling properties were much superior to those of the anhydrous compound. A candidate for entry to the list is ethylene di-hydrazine $(H_3N_2C_2H_4N_2H_3)$ synthesized by Dow early in 1962. By itself it wouldn't be particularly useful—its freezing point is 12.8°C—but its density is high (1.09), and it might well be superior to DETA as a density additive.

So now the designer has a family of high performing fuels at his disposal—reliable, easy to handle, and available. Which mixture he chooses—or composes for the occasion—depends upon the specific requirements of the job at hand. And he knows that it will work. That, at least, is progress.

4

. . . and Its Mate

The RFNA of 1945 was hated by everybody who had anything to do with it, with a pure and abiding hatred. And with reason. In the first place, it was fantastically corrosive. If you kept it in an aluminum drum, apparently nothing in particular happened—as long as the weather was warm. But when it cooled down, a slimy, gelatinous, white precipitate would appear and settle slowly to the bottom of the drum. This sludge was just sticky enough to plug up the injector of the motor when you tried to fire it. People surmised that it was some sort of a solvated aluminum nitrate, but the aversion with which it was regarded was equaled only by the difficulty of analyzing it.

If you tried to keep the acid in stainless steel (SS-347 stood up the best) the results were even worse. Corrosion was faster than with aluminum, and the acid turned a ghastly green color and its performance was seriously degraded. This became understandable when the magnitude of the change in composition was discovered. Near the end of 1947, JPL published the results of two acid analyses. One was of a sample of RFNA fresh from the manufacturer, which had scarcely started to chew on the drum in which it was shipped. The other was a sample of "old" acid, which had been standing for several months in a SS-347 drum. The results were eloquent. And, if my own experience is any criterion, there was a bit of insoluble matter of cryptic composition on the bottom of the drum. Acid like that might have been useful in the manufacture of fertilizer, but as a propellant it was not.[*]

---

[*] Note to the sophisticated reader: Don't take the exact percentages too seriously. Acid analysis wasn't really that good in 1947. Also, most of the iron really shows up in the ferrous and not in the ferric state, as I discovered in my own laboratory (and to my complete surprise) some years later.

Constituent	New acid	Old acid
HNO_3	92.6 percent	73.6 percent
N_2O_4	6.3 percent	11.77 percent
$Fe(NO_3)_3$	.19 percent	8.77 percent
$Cr(NO_3)_3$	.05 percent	2.31 percent
$Ni(NO_3)_2$	.02 percent	.71 percent
H_2O	.83 percent	2.83 percent

So the acid couldn't be kept indefinitely in a missile tank—or there wouldn't be any tank left. It had to be loaded just before firing, which meant handling it in the field.

This is emphatically not fun. RFNA attacks skin and flesh with the avidity of a school of piranhas. (One drop of it on my arm gave me a scar which I still bear more than fifteen years later.) And when it is poured, it gives off dense clouds of NO_2, which is a remarkably toxic gas. A man gets a good breath of it, and coughs a few minutes, and then insists that he's all right. And the next day, walking about, he's just as likely as not to drop dead.

So the propellant handlers had to wear protective suits (which are infernally hot and so awkward that they probably cause more accidents than they prevent) and face shields, and frequently gas masks or self-contained breathing apparatus.

An alternative to RFNA was mixed acid, essentially WFNA to which had been added some 10 to 17 percent of H_2SO_4. Its performance was somewhat lower than that of RFNA (all that stable sulfuric acid and that heavy sulfur atom didn't help any) but its density was a little better than that of the other acid, and it was magnificently hypergolic with many fuels. (I used to take advantage of this property when somebody came into my lab looking for a job. At an inconspicuous signal, one of my henchmen would drop the finger of an old rubber glove into a flask containing about 100 cc of mixed acid—and then stand back. The rubber would swell and squirm a moment, and then a magnificent rocket-like jet of flame would rise from the flask, with appropriate hissing noises. I could usually tell from the candidate's demeanor whether he had the sort of nervous system desirable in a propellant chemist.) Mixed acid, of course, didn't give off those NO_2 fumes, and everybody was convinced, as late as 1949, that it didn't corrode stainless steel. In that year the Navy purchased several hundred 55-gallon drums and several tank cars, all expensively (the drums cost about $120 each) made from SS-347, and designed to contain mixed acid.

Well, everybody had been wrong. The acid doesn't corrode stainless—at first. But after an induction period, which may vary from minutes to months, and which depends upon the acid composition and particularly the percentage of water, the temperature, the past history of the steel, and presumably upon the state of the moon, the corrosion starts and proceeds apace. The eventual results are worse than with RFNA. Not only is the quality of the acid degraded and the drum damaged, but a thick, heavy, greenish-gray sludge of loathsome appearance, revolting properties, and mysterious composition forms and deposits. I have seen drums of mixed acid with twelve solid inches of sludge on the bottom. To make things worse, pressure gradually builds up in the drum or tank car, which has to be vented periodically. And the water breathed in then (mixed acid is extremely hygroscopic) accelerates the corrosion. Within two years all the Navy's expensive tank cars and drums had to be junked.

Another possibility was white fuming nitric acid, which, at least, didn't give off lethal clouds of NO_2 when it was poured. But its freezing point was too high to be acceptable. (Pure HNO_3 freezes at $-41.6°$, the commercial WFNA a few degrees lower.) It was just as corrosive as RFNA, if not more so, and was less hypergolic with many fuels than the red acid. And it had another trick up its sleeve. For years people had noted that a standing drum of acid slowly built up pressure, and had to be vented periodically. But they assumed that this pressure was a by-product of drum corrosion, and didn't think much about it. But then, around the beginning of 1950, they began to get suspicious. They put WFNA in glass containers and in the dark (to prevent any photochemical reaction from complicating the results) and found, to their dismay, that the pressure buildup was even faster than in an aluminum drum. Nitric acid, or WFNA at least, was inherently unstable, and would decompose spontaneously, all by itself. This was a revolting situation.

The fourth possibility was N_2O_4. True, it was poisonous, but if you could avoid handling it in the field that didn't much matter. And, as long as you kept water out of it, it was practically noncorrosive to most metals. You didn't even have to keep it in aluminum or stainless—ordinary mild steel would do. So the tanks of a missile could be filled at the factory, and the operators would never see, or smell, or breathe, the N_2O_4. And it was perfectly stable in storage, and didn't build up any pressure. But its freezing point was $-9.3°$, which the services would not accept.

Thus, with four oxidizers available, we had four sets of headaches—and nothing that we could use with any degree of satisfaction. The situation led to what might be called "the battle of the acid," which went on for some five years, and involved just about every chemist in the rocket business—and a lot who were not.

There were certainly problems enough for everybody, more than enough to go around. As a result, research went off in a dozen different, and at times contradictory, directions. Several groups attacked the freezing point of WFNA directly, using all sorts of additives to bring it down to a reasonable (or, in the case of those shooting for $-100°F$, an unreasonable) figure. R. W. Greenwood at Bell Aircraft, and R. O. Miller, of the Lewis Flight Propulsion Laboratory of NACA, both investigated ammonium nitrate and a 50 percent aqueous solution of the salt; 72 percent perchloric acid (the anhydrous stuff was entirely too touchy to handle) and a 50 percent solution of potassium nitrate (the dry salt was almost insoluble in WFNA), which had been suggested by WADC. They got their freezing points down where they wanted them, but at an intolerable cost. Ignition in a motor was slow and frequently explosive, and combustion was rough and unsatisfactory. And the KNO_3 solution had another disadvantage, which had not been anticipated. When it was fired, the exhaust stream contained a high concentration of potassium ions and free electrons—a plasma, in fact –which would absorb radio waves like mad and make radar guidance of a missile quite impossible. Greenwood tried a few organic additives, acetic anhydride and 2,4,6 trinitrophenol among them, but that approach was a blind alley. Nitric acid *does* react with acetic anhydride in time—and as for the trinitrophenol, loading a propellant up with a high explosive isn't a very appealing idea.

W. H. Schechter, of the Callery Chemical Co., with more courage than judgment, investigated anhydrous perchloric acid, but found that he couldn't get the depression he wanted with a percentage of the additive that could be lived with, and also tried nitronium perchlorate. He didn't get any freezing point depression to speak of, the stability of the mixture was worse than that of the straight WFNA, and its corrosivity was absolutely ferocious. One other additive that he tried was nitromethane, as did A. Zletz, of the Standard Oil Company of Indiana, who also investigated the ethyl and 2 propyl homologues. Nitromethane, naturally, was the best depressant of the lot, and a freezing point of $-100°F$ was reached without any trouble, but the mixture was too sensitive and likely to explode to be of any use.

Mike Pino, of California Research, tried sodium nitrite (it worked, but slowly reacted with the acid to form sodium nitrate, which precipitated out) and sodium cobaltinitrite and found that 4 percent of the salt plus 1 percent of water reduced the freezing point of *anhydrous* acid to $-65°F$, but he couldn't get to the magic $-100°F$ with any reasonable amount of water. He was always very conscious of the effect (pernicious) of water on ignition delay, and shied away from any system that contained any great amount of it. The mixture was unstable, too. So he took another tack, and went to work to see if he could do anything with mixed acid. He had already tried nitrosyl sulfuric acid, $NOHSO_4$, and had found that it was a better freezing point depressant than sulfuric acid, but that it was even worse as a sludge producer. He turned then to the alkane sulfonic acids, particularly methane sulfonic acid, and found that 16 percent of this in WFNA gave a mixture that froze only at $-59°$, although upon occasion it could be supercooled considerably below that before solidifying. This looked promising. It gave good ignition with the fuels he was considering at the time (mixtures of allyl amines and triethylamine). Its corrosivity was similar to or a little less than that of WFNA or of ordinary mixed acid, but it had one shining virtue—it didn't produce any sludge. A similar mixed acid was investigated at North American Aviation at about the same time (1953). This used fluorosulfonic acid instead of the methanesulfonic, and most of its properties were very similar to those of the other mixture. But by this time nobody cared.

Many people were more interested in the ignition delay of WFNA than in its freezing point, and they tried to get the driest acid that could be got, in order to determine, exactly, the effect of water on the delay. The General Chemical division of Allied Chemical and Dye Co. could, and would, oblige. Apparently one of their acid stills was unusually efficient, and would turn out acid with less than 1 percent water in it. You could get it, on special order, shipped in 14-gallon glass carboys inside a protective aluminum drum. When it arrived, it was advisable to keep the carboy in a cold box—the colder the better—to slow down the decomposition of the acid.

The work with this "anhydrous" acid extinguished any remaining doubt that ignition delay with WFNA was critically and overwhelmingly dependent on its water content. Nothing else really mattered.

It had become painfully obvious that you had to know how much water you had in your acid before you could load it into a missile and push the button without disaster. It was equally obvious that setting up an analytical chemistry laboratory in the field wasn't practical politics. So a great cry went out for a "field method" for analyzing nitric acid. What the customer wanted, of course, was a little black box into which he could insert a sample of the acid in question (or preferably, that he could merely point at the sample) where upon the box would flash a green light if the acid could be used, or a red one if it couldn't.

Little black boxes like that aren't too easy to come by. But two people tried to invent such a gadget.

The first was Dr. L. White, of the Southern Research Institute, working for the Air Force. His idea was simple and direct. Water, dissolved in nitric acid, has an absorption line in the near infrared. You merely shine IR of the correct wave length through your sample, measure the absorbtion, and there you are. (Another IR absorption band could be used to measure the N_2O_4 content.) Neat, simple—any rocket mechanic can do it.

But things didn't turn out that way. There were the expected difficulties (only they were worse than expected) that stemmed from the corrosive nature of the acid and its fumes, both of which did their best to chew up the black box. But then something much more disconcerting showed up. White would take a sample of acid which was, as far as he could tell, absolutely anhydrous, with no water in it at all. And the IR absorption band was still there, as large as life, and twice as natural. Nitric acid appeared to be a somewhat more complicated substance than most people thought.

It is. Take 100 percent nitric acid—pure hydrogen nitrate. (I won't go into the question of how you go about getting such a substance.) Does it appear as HNO_3, period? It does nothing of the sort. Studies by Ingold and Hughes, by Dunning, and by others during the 30's and 40's had shown that there is an equilibrium:

$$2HNO_3 \rightleftharpoons NO_2^+ + NO_3^- + H_2O,$$

so that there is some—not much, but some—"species" water present even in absolutely "anhydrous" acid. So the relation between "analytical" water, which was what people were interested in, and optical absorption is not linear, and you have to analyze dozens of samples of acid in order to establish a calibration curve. White embarked upon the calibration.

At NARTS, working for the Navy, I was the other black-box builder. I based my method on the electrical conductivity of the acid. If you take pure water and start adding nitric acid to it, queer things happen. The conductivity increases at first, from the practically zero conductivity of pure water, to react a broad maximum at about 33 percent acid. Then it declines, reaching a minimum at about 97.5 percent acid, and then starts to rise again and is still increasing when you get to 100 percent HNO_3. To make the whole thing more complicated, the presence of N_2O_4 in the acid changes the conductivity, too, since N_2O_4 is partially ionized to NO^+ and NO_3^-.

After blundering about a bit, in the spring of 1951 I took the following approach: I would divide a specimen of acid into three parts. Part 1 was left alone. To part 2 I added a small amount of water, 2.5 cc to 50 cc of acid. Part 3 was diluted more liberally, 30 cc of water to 10 of acid. I then measured the conductivities of all three parts and derived two ratios: conductance 1:conductance 2, and conductance 2:conductance 3. (Taking these ratios eliminated the conductivity cell-constant and reduced the effect of temperature variations.) The water and N_2O_4 content of the acid could then, in principle, be deduced from the two ratios. *After,* of course, the method had been calibrated, by measuring the conductivities of 150 or so samples of acid of varying but *known* composition.

And how do you get to know the composition of an acid? By analyzing it, of course. Everybody knows that. So it was something of a shock to the black-box builders to learn that nobody could analyze nitric acid accurately enough to calibrate the field methods.

Obviously, a calibration method has to be better than the method calibrated—and nobody could determine the water content of nitric acid—routinely—to a tenth of a percent. The N_2O_4 was easy—titration with ceric sulfate was fast and accurate. But there was no direct method for determining the water. You had to determine the total acid (HNO_3 plus N_2O_4) and then determine the N_2O_4, and then get the water by difference—a small difference between two large quantities.

Suppose that your analysis said that you had 0.76 percent N_2O_4, and 99.2 percent, plus or minus 0.2 percent, nitric acid (and it was a good man who could be sure of the acid to 0.2 percent!), then what was your water content? 0.04 percent? Minus 0.16 percent? 0.24 percent? You could take your choice—one guess was as good as another.

Many attempts, all unsuccessful, were made to find a direct method for water, but I chose to apply brute force, and set out grimly to refine the classical method until it could be used to calibrate the field methods. Every conceivable source of error was investigated—and it was surprising to learn in how many ways a classical acid-base titration can go wrong. Nobody would have believed, until he learned the hard way, that when you make up five gallons of 1.4 normal NaOH, you have to stir the solution for an hour to make sure that its concentration is uniform to within one part in 10,000 throughout the whole volume. Nor that when air is admitted to the stock bottle it has to be bubbled through a trap of the same solution. If it isn't, the moisture in the laboratory air will dilute the upper layer of the NaOH and foul you up. Nor that when you get to a phenolphthalein end-point with your 1.4 N alkali, it's advisable to back-titrate with 0.1 N HCl (thus splitting the last drop) until the pink color is the faintest discernible tint. But all those precautions and refinements are necessary if you need results that you can believe.

The most important refinement was the use of specially made precision burettes, thermostated and held at 25°. (The coefficient of expansion of 1.4 N NaOH was *not* well known, and even if it were, somebody would be sure to put it in backwards!) The burettes were made for me by the Emil Greiner Co., and cost the taxpayer seventy-five dollars a throw. They worked so well that certain other agencies acquired the deplorable habit of borrowing one from me and then forgetting to return it.[*]

The job took almost a year, but when it was done the water in the acid could be determined, by *difference,* to 0.025 percent. And the analysis took no longer than the crude analysis of a year before.

The calibration then went like a breeze, complicated only by the difficulties encountered when absolutely anhydrous acid was needed. The classical way of making such a substance was to mix P_2O_5 with WFNA, and then distill the dry acid over under vacuum. This was an infernal nuisance—three hours work might get you ten cc of anhydrous acid—and in our case we needed it by the liter. So we hit on a simple method that required no effort or attention whatsoever. Into a big flask we would load about two liters of 100 percent sulfuric

---

[*] I name no names, but God will punish Doc Harris of WADC!

acid, and then three times as much WFNA. Then, holding the flask at about 40°, we would blow dry air through it, and try to condense as much acid as we could out of the exhaust stream. We'd start the gadget going in the evening, and by next morning there would be a liter or two of water-white acid (the N_2O_4 had all been blown out) waiting to be stored in the deep freeze. It would analyze from 99.8 percent to more than 100 percent acid—the last, of course, containing excess N_2O_5. The method was horribly inefficient—we lost two-thirds of the acid in the exhaust—but with acid at nine cents a pound, who cared?

White published his complete optical method for water and N_2O_4 at the end of 1951, and I published my conductivity method nine months later.[*] Both black boxes worked fine. And, then, naturally, everybody lost interest in WFNA.

There were a few other analytical problems connected with nitric acid that were cleaned up at about this time. Dr. Harris, at WADC, designed an ingenious glass and Teflon sample holder for RFNA, which made it possible to prevent any loss of N_2O_4 when the acid was diluted before titration, and let it be analyzed with an accuracy equal to that possible with WFNA. And I devised analyses for mixed acid and for Mike Pino's mixture of WFNA and methane sulfonic acid. These are worth recording, if only to show the weird expedients to which we were driven to get the results we needed. In both cases, the N_2O_4 and the total acid were determined exactly as in the refined WFNA analysis, and the problem was to determine the additive acid. In the case of the mixed acid, the major part of the nitric acid in the sample was destroyed with formaldehyde, and any formic acid formed was reacted with methanol and boiled off as methyl formate. (The emerging fumes invariably caught fire and burned with a spectacular blue flame.) What was left, then, was dumped into a boiling mixture of water and n-propanol, and titrated, conductimetrically, with barium acetate. This sounds like a weird procedure, but it worked beautifully, and gave as precise results as anybody could wish. Mike Pino's mixture had to be treated differently. The nitric acid was destroyed by reacting it with warm formic acid, and what was left was titrated, potentiometrically, with sodium acetate in acetic acid, in a medium of glacial acetic acid. One electrode was a conventional glass electrode as used for pH determination, the other a modified calomel electrode, using saturated lithium chloride in acetic

---

[*] Dave Mason and his associates at JPL, about sixteen months later, in January 1954, described another conductimetric method, which would work with both WFNA and RFNA. Two conductivity measurements were made, both at 0°C—one of the straight acid and one of the acid saturated with KNO_3. From these two measurements the N_2O_4 and H_2O could be derived using a calibration chart.

acid. Again, a peculiar but effective analysis. And as soon as these methods had been worked out, everybody stopped using either mixed acid!

In many ways N_2O_4 was more appealing as an oxidizer than nitric acid. Its performance was a little better, and it didn't have so many corrosion problems. Its main drawback, of course, was its freezing point, and several agencies tried to do something about that. The prime candidate for a freezing point depressant was nitric oxide, NO. Wittorf, as early as 1905, had examined the phase behavior of the mixture, as had Baumé and Roberts in 1919. But mixtures of NO and N_2O_4 have a higher vapor pressure than the neat nitrogen tetroxide, and several optmists tried to find an additive that would reduce the freezing point without increasing the vapor pressure. This turned out to be rather easy to do—lots of things are soluble in N_2O_4—but at an unacceptable price. L. G. Cole, at JPL, in 1948, tried such things as mono and di nitrobenzene, picric acid, and methyl nitrate, and discovered, upon examining his mixtures, that he had some extremely touchy and temperamental high explosives on his hands. T. L. Thompson, at North American, three years later, tried nitromethane, nitroethane, and nitropropane, and made the same discovery. Collins, Lewis, and Schechter, at Callery Chemical Co., tried these same nitro-alkanes in 1953, as well as tetranitromethane, and worked out the ternary phase diagram for nitrogen tetroxide, nitromethane, and TNM.

Again—high explosives. At about the same time, S. Burket, at Aerojet, went them one better by trying not only these compounds, but even the notoriously treacherous nitroform, plus diethyl carbonate, diethyl oxalate, and diethyl cellosolve. And his mixtures, too, were nothing more than catastrophes looking for a place to happen. It appeared that about the only thing that could safely be dissolved in a nitrogen oxide was another nitrogen oxide.

T. L. Thompson had tried nitrous oxide in 1951, and reported that it wasn't particularly soluble in N_2O_4, and this was confirmed by W. W. Rocker of du Pont. So nitric oxide it had to be.[*]

NO is an extremely effective freezing point depressant for N_2O_4. It combines, under pressure or at low temperatures, with the latter to form the unstable N_2O_3, so that the eutectic appears between pure N_2O_4 and the composition corresponding to N_2O_3, so that a small addition of NO has an inordinately large effect on the freezing point. G. R. Makepeace and his associates, at NOTS, were able to show, in 1948, that 25 percent of NO would bring the freezing point of nitrogen tetroxide down below the required −65°F, and that 30 percent would depress it well below the magic −100°F. However, the vapor

---

[*] Cole, at JPL, had reported in 1948 that a mixture of 41.5 percent N_2O and the remainder N_2O_4 had a freezing point of −51° and a boiling point of 33°. These figures so thoroughly contradicted the experience of everybody else that they are completely inexplicable.

pressure of the latter mixture at 160°F was unacceptably high, about 300 psi. Several investigators examined the system, among them T. L. Thompson of North American and T. J. McGonnigle of, appropriately, the Nitrogen Division of Allied Chemical and Dye Co., but the definitive work came from JPL and NOTS.

Between 1950 and 1954, Whittaker, Sprague, and Skolnik and their group at NOTS, and B. H. Sage and his colleagues at JPL investigated the nitrogen tetroxide–nitric oxide system with a thoroughness that left nothing to be discovered that could conceivably be worth the trouble of discovering. Their meticulous investigations were to bear fruit years later, when Titan II, with its N_2O_4 oxidizer, was developed.

Several agencies tried the mixed oxides of nitrogen (MON-25 or MON-30 or whatever, with the number designating the percentage of NO in the mix) with various fuels, and discovered that it was more difficult to get a good performance (a high percentage of the theoretical performance) with MON than with neat nitrogen tetroxide. Apparently the great kinetic stability of the NO slowed down the combustion reaction. For this reason, and because of its high vapor pressure, investigators turned away from MON for some years. (Certain space rockets, today, use MON-10.)[*]

And there was another reason. RFNA had been domesticated. Two things had done it: A series of meticulous studies at Ohio State University and at JPL solved the problem of decomposition and pressure buildup, and a completely unexpected breakthrough at NARTS reduced the corrosion problem to negligible proportions. With these problems solved the acid could be "packaged" or loaded into a missile at the factory, so that it didn't have to be handled in the field. And that solved the problem of those toxic fumes, and eliminated the danger of acid burns.

By the beginning of 1951 the nature and behavior of nitric acid had become comprehensible. True, it was a fiendishly complicated system—one could hardly call it a substance—but some sense could be made out of it. The monumental work of Professor C. K. Ingold and his colleagues, published in a series of articles in 1950, had clarified the equilibria existing among the various species present in the system, and Frank and Schirmer, in Germany, in the same year, explained its decomposition. Briefly, this is what their work showed:

First, in very strong nitric acid, there is an equilibrium:

$$(1) \qquad 2HNO_3 \rightleftharpoons H_2NO_3^+ + NO_3^-.$$

---

[*] And "green" N_2O_4, containing about 0.6% of NO and green by transmitted light, has recently been developed. The NO seems to reduce stress corrosion of titanium, and also scavenges dissolved oxygen in the N_2O_4.

However the concentration of $H_2NO_3^+$ is extremely small at any time, since it, too is in equilibrium:

(2) $$H_2NO_3^+ \rightleftarrows H_2O + NO_2^+,$$

so that for all practical purposes we can write:

(3) $$2HNO_3 \rightleftarrows NO_2^+ + NO_3^- + H_2O$$

and ignore the $H_2NO_3^+$. In dilute acid, the equilibrium is

(4) $$H_2O + HNO_3 \rightleftarrows H_3O^+ + NO_3^-.$$

Thus, in acid containing less than about 2.5 percent of water, NO_2^+ is the major cation, and in acid containing more than that, H_3O^+ takes that role. Exactly *at* 2.5 percent water, very little of either one is present, which very neatly explains the minimum in the electrical conductivity observed there. If NO_2^+ is the active oxidizing ion in strong acid (and in the course of some corrosion studies I made a couple of years later I proved that it is) the effect of water on ignition delay is obvious. Equation (3) shows that adding water to dry acid will reduce the concentration of NO_2^+ which is the active species. The addition of NO_3^- will do the same thing—which explains the poor combustion observed with acid containing NH_4NO_3.

The nitronium (NO_2^+) ion would naturally be attracted to a negative site on a fuel molecule, such as the concentration of electrons at a double or triple bond—which goes far to justify Lou Rapp's remarks as to the desirability of multiple bonds to shorten ignition delay.

The ion also explains the instability of nitrites in strong acid by the reaction:

$$NO_2^- + NO_2^+ \longrightarrow N_2O_4$$

If N_2O_4 is present in strong acid, another set of equilibria show up.

(5) $$2NO_2 \rightleftarrows N_2O_4 \rightleftarrows NO^+ + NO_3^-$$

The result of all of this is that (even neglecting solvation) in strong acid containing N_2O_4 have appreciable quantities of at least seven species:

HNO_3	NO_2^+
N_2O_4	NO^+
NO_2	and
H_2O	NO_3^-

Plus possible traces of H_3O^+ and $H_2NO_3^+$. And all of them in inter-locking equilibria. But this didn't explain the pressure buildup. Nitric acid decomposes by the gross reaction.

(6) $$4HNO_3 \longrightarrow 2N_2O_4 + 2H_2O + O_2$$

But how? Well, Frank and Schirmer had shown that there is yet another equilibrium present in the system, and another species:

$$(7) \qquad\qquad NO_3^- + NO_2^+ \rightleftarrows N_2O_5$$

And N_2O_5 was well known to be unstable and to decompose by the reaction.

$$(8) \qquad\qquad N_2O_5 \longrightarrow N_2O_4 + \frac{1}{2}O_2$$

Then as O_2 is essentially insoluble in nitric acid, it bubbles out of it and the pressure builds up and your acid turns red from the NO_2.

What to do about it? There were two possible approaches. The obvious one is suggested by equation (6): increase the concentration (or, in the case of the oxygen, the pressure) of the species on the right hand side of the equation, and force the equilibrium back. It soon became obvious that merely putting a blanket of oxygen over your WFNA wouldn't help. The equilibrium oxygen pressure was much too high. I have actually seen the hair-raising sight of rocket mechanics trying to determine the oxygen pressure developed over decomposing WFNA by measuring the bulging of the drums—and shuddered at the sight! The equilibrium oxygen pressure over 100 percent acid at zero ullage (no appreciable unfilled volume in the tank) at 160°F turned out to be well over 70 atmospheres. Nobody wants to work with a bomb like that.

To reduce the equilibrium oxygen pressure, you obviously have to increase the N_2O_4 or the water concentration or both. WFNA and anhydrous acid were definitely out.

It was D. M. Mason and his crew at JPL and Kay and his group at Ohio State who undertook—and completed—the heroic task of mapping the phase behavior and equilibrium pressure and composition of the nitric acid-N_2O_4-H_2O system over the whole composition range of interest, up to 50% N_2O_4 and up to 10 percent or so H_2O—and from room temperature up to 120°C. By the time these groups were finished (all of the work was published by 1955) there was nothing worth knowing about nitric acid that hadn't been nailed down. Thermodynamics, decomposition, ionetics, phase properties, transport properties, the works. Considering the difficulties involved in working with such a miserable substance, the achievement can fairly be classified as heroic.

And it paid off. An RFNA could be concocted which had a quite tolerable decomposition pressure (considerably less than 100 psi) even at 160°F (71°C). The General Chemical Co. came up with one containing 23% N_2O_4 and 2% H_2O, while the JPL mixture, which they called SFNA (Stable Fuming Nitric Acid) contained 14 percent and 2.5 percent respectively.

The freezing points of the HNO_3-N_2O_4-H_2O mixtures were soon mapped out over the whole range of interest. R. O. Miller at LFPL, G. W. Elverum at JPL, and Jack Gordon at WADC among others, were involved in this job, which was completed by 1955.

Their results were not in the best of agreement (the mixtures frequently supercooled and, as I have mentioned, RFNA is not the easiest thing in the world to analyze) but they all showed that both the General Chemical Co. mixture and JPL's SFNA froze below −65°F. About this time the Navy decided to relax and enjoy it and backed off from their demand for the mystic −100°F and everybody and his brother heaved a deep sigh of relief. One job done!

The solution to the corrosion problem turned out to be simple—once we had thought of it. In the spring of 1951 we at NARTS were concerned about—and studying—the corrosion of 18-8 stainless steel, specifically SS-347, by WFNA. Eric Rau, who had been with me for only a few months (the chemistry lab had been functioning only since the previous summer) thought that a coating of fluoride on the steel might protect it from the acid. (Don't ask me why he thought so!) So, he talked a friend of his who worked at the General Chemical Co. division of Allied Chemical and Dye into taking some of our sample strips of 347 and leaving them for some days inside one of the pipelines that conveyed HF from one part of the plant to another. Then Eric tested these samples for corrosion resistance, and found that they corroded just as badly as did the untreated steel. *But,* this corrosion was delayed, and didn't start, apparently, until a day or two had passed. The inference was that (1) a fluoride coating was protective, but (2) it didn't last long in WFNA. He thought then that it might be possible to make the fluoride coating self-healing by putting some HF in the WFNA. However the only HF that we had in the lab was the common 50 percent aqueous solution of that acid, and Eric didn't want to add any water to his WFNA. So I suggested that he try ammonium bifluoride, $NH_4F \cdot HF$, which is more than two-thirds HF anyway, and a lot easier to handle. Also, we had it on the shelf. He tried it, and to our incredulous delight it worked—worked with an effectiveness beyond our wildest hopes. A few weeks of messing around showed us that 0.5 percent of HF in the acid, no matter how introduced, reduced the corrosion rate of the steel by a factor of ten or more, and that more than 0.5 percent didn't improve things measurably. We reported this finding in our quarterly report, on 1 July, 1951, but NARTS was just two years old then, and apparently nobody bothered to read our reports.

But there was a meeting at the Pentagon devoted to the problems of nitric acid on October 10–11–12, attended by about 150 propellant-oriented people from industry, government and the services. I went, and so did Dr. Milton Scheer ("Uncle Milty") of our group, and on the afternoon of the 11th he reported Eric's discovery. What made the occasion delightful (for us) was the fact that that very morning, in discussing another paper, R. W. Greenwood, of Bell Aircraft, had stated that he had tried ammonium bifluoride as a freezing point depressant for WFNA, and then, three papers later, T. L. Thompson of North American Aviation reported on using both anhydrous and aqueous HF as freezing point depressants for R.F.N.A. And both of them had completely missed the corrosion-inhibiting effect!

Then everybody got into the act—North American, JPL, and just about everybody else. (We were already there.) As it turned out, HF was even more effective in inhibiting the corrosion of aluminum than reducing that of SS-347: inhibition was just as good with RFNA as with WFNA; and it was effective not only in the liquid phase but in the gas phase, where the metal was in the acid vapor above the liquid level.

But while HF was a good inhibitor for aluminum and for 18-8 stainless steels, it wasn't universally effective. It had no particular effect on the corrosion of nickel or chromium, while it *increased* the corrosion rate at tantalum by a factor of 2000 and that of titanium by one of 8000.

There was a great deal of interest in titanium at that time, and as many rocket engineers wanted to use it, the question of its resistance to RFNA couldn't be neglected. But these corrosion studies were interrupted by a completely unexpected accident. On December 29, 1953, a technician at Edwards Air Force Base was examining a set of titanium samples immersed in RFNA, when, absolutely without warning, one or more of them detonated, smashing him up, spraying him with acid and flying glass, and filling the room with NO_2. The technician, probably fortunately for him, died of asphyxiation without regaining consciousness.

There was a terrific brouhaha, as might be expected, and JPL undertook to find out what had happened. J. B. Rittenhouse and his associates tracked the facts down, and by 1956 they were fairly clear. Initial intergranular corrosion produced a fine black powder of (mainly) metallic titanium. And this, when wet with nitric acid, was as sensitive as nitroglycerine or mercury fulminate. (The driving reaction, of course, was the formation of TiO_2.) Not all titanium alloys behaved this way, but enough did to keep the metal in the doghouse for years, as far as the propellant people were concerned.

In spite of the titanium debacle, the rocket business now had a usable nitric acid, and a rewriting of the military specifications for WFNA and RFNA seemed appropriate.

During 1954, then, a group representing the services and industry got together under Air Force sponsorship to do just that. I was there, as one of the Navy representatives.

Various users still argued over the relative merits of 14 percent RFNA and 22 percent RFNA, and a few still liked WFNA. The chemical industry was amiably willing to go along with anything—"Hell, it's just as easy to make one sort of acid as another—just tell us what you want!" So we decided to write one specification which would make everybody happy. We officially threw out the terms WFNA and RFNA and described no less than four types of nitric acid, which we designated, with stunning unoriginality as "Nitric Acid, Type I, II, III and IV." These contained, in the order named, nominally 0 percent, −7 percent, 14 percent, and 21 percent N_2O_4. If you wanted HF inhibited acid, you asked for I-A or III-A, or whatever, and your acid would contain 0.6 percent HF.

I was against describing the nature of the inhibitor in the openly published specifications, since the inhibition was such an unlikely—though simple—trick that it might well have been kept secret for some time. I had friends in the intelligence community, and asked them to try to learn, discreetly, whether or not the trick was known on the other side of the iron curtain. The answer came back, with remarkable speed, that it was not, and that, in fact, the Soviet HF manufacture was in trouble, and that the director of the same was vacationing in Siberia. So I protested violently and at length, but the Air Force was running the show and I was overruled. And when the specs were published, the gaff was blown for good.

Included in the specs were the procedures for analyzing the acids. These were conventional, except the one for HF, which was a complicated and tricky optical method involving the bleaching of a zirconium-alizarin dye by fluoride ion. In my own lab I declined to have anything to do with it, and whomped up a simple—not to say simple-minded—test that required no effort or intelligence whatever. You put one volume of acid and two of water in a polyethylene beaker, and dropped into it a magnetic stirring rod enclosed in soft glass tubing and weighed. You then let the thing stir overnight and reweighed the stirring rod. If you had calibrated that particular piece of glass with an acid containing a known concentration of HF, that was all you needed. Accuracy quite good enough for the purpose.

Dave Mason of JPL came up with another quick-and-dirty method for estimating the HF—almost as simple as mine, and a lot faster. It was a colorimetric method, which depended upon the bleaching effect of fluoride ion on purple ferric salicylate.

As it turned out, the type III-A gradually edged out the others, and is now *the* nitric acid oxidizer.* The engineers call it IRFNA, inhibited Red Fuming Nitric Acid, and very few of the current crop are even aware that there ever was another sort—or of what "inhibited" means. A few years ago I saw one alleged rocket engineer fill a stainless steel tank with RFNA *without* any HF in it—and then wonder why his acid turned green.

---

* Just one important motor—that for the second stages of Vanguard and of Thor Able used type I-A acid (IWFNA) which it burned with UDMH.

The only other sort of acid worth mentioning is "Maximum Density Nitric Acid." This was proposed by Aerojet for applications in which density is all-important and freezing point requirements are not too strict. It contains 44 percent N_2O_4 and has a density of 1.63. Once a satisfactory acid had been found, interest in its analysis dropped to zero. III-A was so smoothly hypergolic with UDMH, and a little water more or less didn't make any difference, and you could keep it sealed so it wouldn't pick up water—and with the HF in it there wasn't any corrosion to worry about—so why bother? An occasional purchasing agent may have a drum analyzed now and then, but the general custom is to accept the manufacturer's analysis—slap the acid into the tank—and fire it. And it works.

The situation today, then, is this: For tactical missiles, where the freezing point of the propellants matters, IRFNA type III-A is the oxidizer. The 47,000-pound thrust Lance, whose fuel is UDMH, is an example, as is the Bullpup, which burns a mixture of UDMH, DETA and acetonitrile. In space, Bell's remarkably reliable Agena motor, of 16,000 pounds of thrust, also uses IRFNA, along with UDMH.

For strategic missiles, which are fired from hardened—and heated—sites, N_2O_4, with a somewhat greater performance, is the oxidizer used. Titan II is, of course, the largest of the US ICBMS, and its first stage is driven by two 215,000-pound thrust motors, using N_2O_4 and the 50–50 hydrazine-UDMH mixture.

Many other N_2O_4 motors are used in space, ranging from the 21,500-pound Apollo service engine, which also uses 50–50, down to tiny one-pound thrusters used for attitude control. The fuel is invariably a hydrazine or a hydrazine mixture. And the users have reason to be happy with their performance and reliability.

As have the chemists, and engineers, who don't have to go through it again.

Afterword

Another symposium on liquid propellants was held at the Pentagon on May 23 and 24, 1955. If the October 1951 meeting was devoted mainly to difficulties, the May 1955 meeting described a series of battles fought and triumphantly won.

The high points were the narration by Bernard Hornstein of ONR of the development of MMH and UDMH, and that by S. P. Greenfield, of North American, of the vicissitudes of NALAR.

NALAR was a 2.75″ diameter air-to-air missile for the Air Force. The requirements were rough. The liquid propellants had to be hypergolic. They also had to be packageable, so that the missile could be stored, fully fueled, for five years and be in a condition to fire. And they had to perform at any temperature from −65°F to +165°F. North American started development in July 1950.

The first oxidizer they tried was RFNA, 18% N_2O_4. From the beginning they were contending with a pressure buildup, and with corrosion. However, trying to get good ignition and smooth combustion, they fired it with:

	Turpentine
and	Decalin
and	2Nitropropane plus 10–20% turpentine
and	Isopropanol
and	Ethanol
and	Butylmercaptan
and	Toluene
and	Alkyl thiophosphites
and	got nowhere.

Then they shifted to MON-30 for their oxidizer, 70% N_2O_4, 30% NO, and resumed their quest for smooth ignition and smooth combustion with:

	Turpentine
and	Butyl mercaptan
and	Hydrazine
and	Isopropanol
and	Toluene
and	2Methyl furan
and	Methanol
and	Aviation gasoline
and	Turpentine plus 20–30% 2Methyl furan
and	Butyl mercaptan plus 20–30% 2Methyl furan
and	Isopropanol plus 30% turpentine
and	Methanol plus 20–25% 2Methyl furan
and	Methanol plus 30–40% Hydrazine
and	Alkyl thiophosphites
and	Turpentine plus Alkyl thiophosphites
and	JP-4 plus Alkyl thiophosphites
and	JP-4 plus 10–30% Xylidine
and	achieved a succession of hard starts, usually followed by rough combustion.

By this time the spring of 1953 had arrived, and the engineers learned of the uses of HF in inhibiting nitric acid corrosion. (The fact that this effect had been discovered two years before, and that North American's own chemists had been working with HF for at least a year suggest that there was a lack of communication somewhere, or, perhaps, that engineers don't read!)

Be that as it may, they returned, probably with a sense of *déja vu,* to turpentine and RFNA—but inhibited this time. To improve ignition they added up to 20 percent of Reference Fuel 208, the alias of 2-di-methylamino-4-methyl-1-3-2-dioxaphospholane, to the turpentine. Then the Air Force, who, you will recollect, was paying for all of this, suggested that they substitute UDMH for the RF-208. They did, and the results were so good that they went to straight UDMH, and to Hell with the turps.

It had taken them four years to arrive at today's standard work-horse combination of UDMH-IRFNA, but they had finally arrived. And recently, a NALAR missile which had been sitting around for about twelve years was hauled off the shelf and fired. And it worked. The hypergol and his mate had been captured and tamed. (Fade out into the sunset to the sound of music.)

5

Peroxide—Always a Bridesmaid

Hydrogen peroxide can be called the oxidizer that never made it. (At least, it hasn't yet.) Not that people weren't interested in it—they were, both in this country and, even more so, in England. Its performance with most fuels was close to that of nitric acid, as was its density, and in certain respects it was superior to the other oxidizer. First, no toxic fumes, and it didn't chew on skin as the acid did. If you received a splash of it, and didn't delay too long about washing it off, all the damage you got was a persistent itch, and skin bleached bone white—to stay that way until replaced by new. And it didn't corrode metals as the acid did.

But (as is usual in the propellant business, there were lots of "buts") the freezing point of 100 percent H_2O_2 was only half a degree below that of water. (Of course, 85 or 90 percent stuff, which was the best available in the 40's, had a better freezing point, but diluting a propellant with an inert, just to improve its freezing point, is not a process that appeals to men interested in propulsion!) And it was unstable.

Hydrogen peroxide decomposes according to the equation $H_2O_2 \longrightarrow H_2O + \frac{1}{2}O_2$, *with the evolution of heat.* Of course, WFNA also decomposed, but *not* exothermically. The difference is crucial: It meant that peroxide decomposition is self-accelerating. Say that you have a tank of peroxide, with no efficient means of sucking heat out of it. Your peroxide starts to decompose for some reason or other. This decomposition produces heat, which warms up the rest of the peroxide, which naturally then starts to decompose faster—producing more heat. And so the faster it goes the faster it goes until the whole thing goes up in a magnificent whoosh or bang as the case may be, spreading superheated steam and hot oxygen all over the landscape.

And a disconcerting number of things could start the decomposition in the first place: most of the transition metals (Fe, Cu, Ag, Co, etc.) and their compounds; many organic compounds (a splash of peroxide on a wool suit can turn the wearer into a flaming torch, suitable for decorating Nero's gardens); ordinary dirt, of ambiguous composition, and universal provenance; OH ions. Name a substance at random, and there's a 50–50 chance (or better) that it will catalyze peroxide decomposition.

There were certain substances, stannates and phosphates, for instance, that could be added to peroxide in trace quantities and would stabilize it a bit by taking certain transition metal ions out of circulation, but their usefulness and potency was strictly limited; and they made trouble when you *wanted* to decompose the stuff catalytically. The only thing to do was to keep the peroxide in a tank made of something that didn't catalyze its decomposition (very pure aluminum was best) and to keep it clean. The cleanliness required was not merely surgical—it was levitical. Merely preparing an aluminum tank to hold peroxide was a project, a diverting ceremonial that could take days. Scrubbing, alkaline washes, acid washes, flushing, passivation with dilute peroxide—it went on and on. And even when it was successfully completed, the peroxide would *still* decompose slowly; not enough to start a runaway chain reaction, but enough to build up an oxygen pressure in a sealed tank, and make packaging impossible. And it is a nerve-wracking experience to put your ear against a propellant tank and hear it go "glub"—long pause—"glub" and so on. After such an experience many people, myself (particularly) included, tended to look dubiously at peroxide and to pass it by on the other side.

Well, early in 1945, we laid our hands on a lot of German peroxide, about 80–85 percent stuff. Some of it went to England. The British were very much interested in it as an oxidizer and in the German manufacturing process. In that same year they fired it in a motor using a solution of calcium permanganate to decompose the peroxide, and with furfural as the fuel, and for several years they worked with it and various (mainly hydrocarbon) fuels.

The rest of it came to this country. However, it contained considerable sodium stannate (as a stabilizer) and was not too suitable for experimental work. So the Navy made a deal with the Buffalo Electrochemical Co., which was just getting into production itself making high-strength peroxide. The Navy turned over most of the German peroxide to Becco, who diluted it down to 2 or 4 percent mouthwash or hair bleach (where the stabilizer was a help) and Becco furnishing the Navy with an equivalent amount of new 90 percent stuff without any stabilizer. And then the Navy distributed this to the various workers in the field.

JPL was one of the first agencies in this country to look at peroxide seriously. From late 1944 through 1948 they worked it out, using 87 percent to 100 percent peroxide, and a variety of fuels, including methanol, kerosene, hydrazine, and ethylene diamine. Only the hydrazine was hypergolic with the peroxide; all the other combinations had to be started with a pyrotechnic igniter. One very odd combination that they investigated during this period was peroxide and nitromethane, either straight or with 35 percent nitroethane or with 30 percent methanol. One oddity was the very low O/F ratio, which ran from 0.1 to 0.5 or so. (With hydrazine as a fuel, it would be about 2.0! The large amount of oxygen in the fuel explains the low O/F.)

Other agencies, MIT and GE and the M. W. Kellogg Co. among them, burned peroxide with hydrazines of various concentrations—from 54 percent up to 100 percent, and Kellogg even tried it with $K_3Cu(CN)_4$ catalyst in the hydrazine, as the Germans had done.

In general, everybody got respectable performances out of peroxide, although there were some difficulties with ignition and with combustion stability, but that freezing point was a tough problem, and most organizations rather lost interest in the oxidizer.

Except the Navy. At just that time the admirals were kicking and screaming and refusing their gold-braided lunches at the thought of bringing nitric acid aboard their beloved carriers; they were also digging in their heels with a determined stubbornness that they hadn't shown since that day when it had first been suggested that steam might be preferable to sail for moving a battleship from point A to point B.

So NOTS was constrained to develop a "nontoxic" propellant system based on hydrogen peroxide and jet fuel, and with acceptable low temperature behavior.

A lot of information was available—on the shelf. Maas and his associates, during the 20's, had investigated hydrogen peroxide up and down and sideways, and had dissolved all sorts of things in it, from salt to sucrose. And many of these things were excellent freezing point depressants: 9.5 percent of ammonia, for instance, formed a eutectic which froze at −40°, and a mixture containing 59 percent froze at −54°. (In between, at 33 percent, was the compound NH_4-OOH, which melted at about 25°.) And one containing 45 percent of methanol froze at −40°. These mixtures, however, had one slight drawback—they were sensitive and violent explosives.

The British, as has been mentioned, were intensely interested in peroxide, and Wiseman, of ERDE (Explosives Research and Development Establishment) at Waltham Abbey, pointed out in 1948 that ammonium nitrate was a good freezing point depressant and didn't make it into a high explosive. So the NOTS team (G. R. Makepeace and G. M. Dyer) mapped out the relevant part of the peroxide–AN–water field, and came up with a mixture that didn't freeze above −54°. It was 55 percent peroxide, 25 percent ammonium nitrate and 20 percent water. They fired it successfully with JP-1 early in 1951, but the performance was not impressive. Other peroxide-AN mixtures were fired by NOTS, and, a little later, by NARTS. In the meantime, L. V. Wisniewski, at Becco, had been adding things like ethylene glycol, diethylene glycol, and tetrahydrofuran to peroxide. These mixtures were designed as monopropellants, but they froze at −40°, and RMI tried them as oxidizers for gasoline and JP-4, with indifferent success. Below +10°C, RMI just couldn't get the mixtures to ignite. Also, they were dangerously explosive.

So, the only low-freezing peroxide mixtures which could be used were those containing ammonium nitrate—and these had serious limitations. One of these was that adding AN to the peroxide increased its instability so much that it was likely to detonate in the injector, and was almost certain to go off, taking the motor with it, if you tried to use it for regenerative cooling.

Ignition of a hydrogen peroxide system, particularly one burning gasoline or jet fuel, was always a problem. In some cases, a solution of calcium permanganate was injected along with the propellants at the start of the run, but this was an awkward complication. In some tests (at MIT) a small amount of catalyst (cobaltons nitrate) was dissolved in the peroxide, but this reduced its stability. The fuel was kerosene with a few percent of o-toluidine. A hypergolic or easily ignited starting slug (generally hydrazine, sometimes containing a catalyst) could lead the fuel. An energetic solid-propellant pyrotechnic igniter was used in some cases. Probably the most reliable, and hence the safest, technique was to decompose part or all of the peroxide in a separate catalyst chamber, lead the hot products into the main chamber, and inject the fuel (and the rest of the oxidizer, if any) there. (A stack of screens made of silver wire was an efficient catalyst array.) NARTS designed and fired a motor which incorporated the catalyst chamber in the main chamber.

Most of the Navy work on peroxide was not directed toward missiles, but toward what was called "super performance" for fighter planes—an auxiliary rocket propulsion unit that could be brought into play to produce a burst of very high speed—so that when a pilot found six Migs breathing down his neck he could hit the panic button and perform the maneuver known as getting the hell out of here. The reason for the jet fuel was clear enough; the pilot already had it aboard, and so only an oxidizer tank had to be added to the plane.

But here an unexpected complication showed up. The peroxide was to be stored aboard airplane carriers in aluminum tanks. And then suddenly it was discovered that trace quantities of chlorides in peroxide made the latter peculiarly corrosive to aluminum. How to keep traces of chloride out of *anything* when you're sitting on an ocean of salt water was a problem whose solution was not entirely obvious.

And there was always the problem of gross pollution. Say that somebody dropped (accidentally or otherwise) a greasy wrench into 10,000 gallons of 90 percent peroxide in the hold of the ship. What would happen—and would the ship survive? This question so worried people that one functionary in the Rocket Branch (safely in Washington) who had apparently been reading Captain Horado Hornblower, wanted us at NARTS to build ourselves a 10,000-gallon tank, fill it up with 90 percent peroxide, and then drop into it—so help me God—one rat. (He didn't specify the sex of the rat.) It was with considerable difficulty that our chief managed to get him to scale his order down to one test tube of peroxide and one quarter inch of rat tail.

Carrier admirals are—with good reason—deadly afraid of fire. That was one of the things they had against acid and a hypergolic fuel.

A broken missile on deck—or any sort of shipboard accident that brought fuel and acid together—would inevitably start a fire. On the other hand, they reasoned that jet fuel wouldn't even mix with peroxide, but would just float on top of it, doing nothing. And if, somehow, it caught fire, it might be possible to put it out—with foam perhaps—without too much trouble.

So, at NARTS we tried it. A few drums of peroxide (about 55 gallons per drum) were poured out into a big pan, a drum or two of JP-4 was floated on top, and the whole thing touched off. The results were unspectacular. The JP burned quietly, with occasional patches of flare or fizz burning. And the fire chief moved in with his men and his foam and put the whole thing out without any fuss. End of exercise.

The Lord had his hands on our heads that day—the firemen, a couple of dozen bystanders, and me.

For when we—and other people—tried it again (fortunately on a smaller scale) the results were different. The jet fuel burns quietly at first, then the flare burning starts coming, and its frequency increases. (That's the time to start running.) Then, as the layer of JP gets thinner, the peroxide underneath gets warmer, and starts to boil and decompose, and the overlying fuel is permeated with oxygen and peroxide vapor. And then the whole shebang detonates, with absolutely shattering violence.

When the big brass saw a demonstration or two, the reaction was "Not on *my* carrier!" and that was that.

The Super-P project was dropped for a variety of reasons, but the pan-burning tests were not entirely without influence on the final decision.

It is amusing to note that when actual tests were made of the effects of a big spill of acid and UDMH, the results weren't so frightening after all. There was a big flare, but the two propellants were so reactive that the bulk liquids could never really mix and explode, but were, rather, driven apart. So the flare was soon over, and plain water—and not much of it, considering—was enough to bring things under control. And so acid-UDMH propelled missiles finally got into the carriers' magazines after all.

But peroxide didn't. Research on it continued for some years, and the British designed and built a rocket-driven plane and a missile or two around the peroxide-JP combination, but that was about all, and for some ten years peroxide, as an oxidizer, has been pretty much out of the picture. (Monopropellant peroxide is another story.)

Higher concentrations (you can buy 98 percent stuff now) have appeared in the last few years, and they appear to be rather more stable than the 90 percent material, but all the drum beating indulged in by the manufacturers hasn't got the bridesmaid into a bridal bed. Peroxide just didn't make it.

6

Halogens and
Politics and
Deep Space

While all of this was going on there were a lot of people who were not convinced that peroxide, or acid, or nitrogen tetroxide was the last word in storable oxidizers, nor that something a bit more potent couldn't be found. An oxygen-based oxidizer is all very well, but it seemed likely that one containing fluorine would pack an impressive wallop. And so everybody started looking around for an easily decomposed fluorine compound that could be used as a storable oxidizer.

"Easily decomposed" is the operative phrase. Most fluorine compounds are pretty final—so final that they can be thought of as the ash of an element which has been burned with fluorine, and are quite useless as propellants. Only when fluorine is combined with nitrogen, or oxygen, or another halogen, can it be considered as available to burn something else. And in 1945 not very many compounds of fluorine with these elements were known.

OF_2 was known, but it was difficult to make and its boiling point was so low that it had to be considered a cryogenic. O_2F_2 had been reported, but was unstable at room temperature. NF_3 was known, but its boiling point was too low for a storable. ONF and O_2NF both had low boiling points, and couldn't be kept liquid at room temperature by any reasonable pressure. It was specified, arbitrarily, a few years later, that a storable propellant must not have a vapor pressure greater than 500 psia at 71° (160°F). Fluorine nitrate and perchlorate, FNO_3 and $FClO_4$, were well known, but both were sensitive and treacherous explosives. Of the latter it had been reported that it frequently detonated "upon heating or cooling; freezing or melting; evaporation or condensation; and sometimes for no apparent reason."

That left the halogen fluorides. IF_5 and IF_7 both melted above $0°C$, and the thought of carrying that heavy iodine atom around was not appealing. BrF was unstable. BrF_3 and BrF_5 were known. If either of these were to be used, the pentafluoride was obviously the better bet, since it carried the more fluorine. ClF was low boiling, and didn't have enough fluorine in it. That left ClF_3, and maybe BrF_5 in a pinch, or when density was all important. (It has a density of 2.466 at 25°.)

And that was it, although JPL in 1947 was dreaming wistfully of such improbabilities as F_2O_7, and the Harshaw Chemical Co. spent a good deal of time and money, in 1949 and 1950, trying to synthesize things like $HClF_6$ and ArF_4,[*] and naturally (as we say now, with 20-20 hindsight) got nowhere. They *did* learn a lot about the synthesis and properties of OF_2.

So ClF_3 it had to be. Otto Ruff had discovered the stuff in 1930 (as he had also discovered the majority of the compounds listed above) and the Germans had done a little work with it during the war, and so quite a lot was known about it. The efflorescence of fluorine chemistry sparked by the Manhattan Project led to studies in this country, and the Oak Ridge people, among others, investigated it exhaustively during the late 40's and early 50's. So it wasn't exactly an unknown quantity when the rocket people started in on it.

Chlorine trifluoride, ClF_3, or "CTF" as the engineers insist on calling it, is a colorless gas, a greenish liquid, or a white solid. It boils at 12° (so that a trivial pressure will keep it liquid at room temperature) and freezes at a convenient −76°. It also has a nice fat density, about 1.81 at room temperature.

It is also quite probably the most vigorous fluorinating agent in existence— much more vigorous than fluorine itself. Gaseous fluorine, of course, is much more dilute than the liquid ClF_3, and liquid fluorine is so cold that its activity is very much reduced.

All this sounds fairly academic and innocuous, but when it is translated into the problem of handling the stuff, the results are horrendous. It is, of course, extremely toxic, but that's the least of the problem. It is hypergolic with every known fuel, and so rapidly hypergolic that no ignition delay has ever been measured. It is also hypergolic with such things as cloth, wood, and test engineers, not to mention asbestos, sand, and water—with which it reacts explosively. It can be kept in some of the ordinary structural metals—steel, copper, aluminum, etc.—because of the formation of a thin film of insoluble metal fluoride which protects the bulk of the metal, just as the invisible coat of

---

[*] It has recently been shown that an argon fluoride, probably ArF_2, *does* exist, but it is unstable except at cryogenic temperatures.

oxide on aluminum keeps it from burning up in the atmosphere. If, however, this coat is melted or scrubbed off, and has no chance to reform, the operator is confronted with the problem of coping with a metal–fluorine fire. For dealing with this situation, I have always recommended a good pair of running shoes. And even if you don't have a fire, the results can be devastating enough when chlorine trifluoride gets loose, as the General Chemical Co. discovered when they had a big spill. Their salesmen were awfully coy about discussing the matter, and it wasn't until I threatened to buy my RFNA from Du Pont that one of them would come across with the details.

It happened at their Shreveport, Louisiana, installation, while they were preparing to ship out, for the first time, a one-ton steel cylinder of CTF. The cylinder had been cooled with dry ice to make it easier to load the material into it, and the cold had apparently embrittled the steel. For as they were maneuvering the cylinder onto a dolly, it split and dumped one ton of chlorine trifluoride onto the floor. It chewed its way through twelve inches of concrete and dug a three-foot hole in the gravel underneath, filled the place with fumes which corroded everything in sight, and, in general, made one hell of a mess. Civil Defense turned out, and started to evacuate the neighborhood, and to put it mildly, there was quite a brouhaha before things quieted down. Miraculously, nobody was killed, but there was one casualty—the man who had been steadying the cylinder when it split. He was found some five hundred feet away, where he had reached Mach 2 and was still picking up speed when he was stopped by a heart attack.

This episode was still in the future when the rocket people started working with CTF, but they nevertheless knew enough to be scared to death, and proceeded with a degree of caution appropriate to dental work on a king cobra. And they never had any reason to regret that caution. The stuff consistently lived up to its reputation.

Bert Abramson of Bell Aircraft fired it in the spring of 1948, using hydrazine as the fuel, NACA and North American followed suit the next year, and in 1951 NARTS burned it with both ammonia and hydrazine.

The results were excellent, but the difficulties were infuriating. Ignition was beautiful—so smooth that it was like turning on a hose. Performance was high—very close to theoretical. And the reaction was so fast that you could burn it in a surprisingly small chamber. But. If your hardware was dirty, and there was a smear of oil or grease somewhere inside a feed line, said feed line would ignite and cleverly reduce itself to ashes. Gaskets and O-rings generally had to be of metal; no organic material could be restrained from ignition. Teflon would stand up under static conditions, but if the CTF flowed over it with any speed at all, it would erode away like so much sugar in hot water, even if it didn't ignite. So joints had to be welded whenever possible, and the welds had to be good. An enclosure of slag in the weld could react and touch off a fire without even trying. So the welds had to be made, and inspected and polished smooth and reinspected, and then all the plumbing had to be cleaned out and passivated before you dared put the CTF into the system. First there was a water flush, and the lines were blown dry with nitrogen. Then came one with ethylene trichloride to catch any traces of oil or grease, followed by another nitrogen blow-down. Then gaseous CTF was introduced into the system, and left there for some hours to catch anything the flushing might have missed, and *then* the liquid chlorine trifluoride could be let into the propellant lines.

It was when the stuff got into the motor that the real difficulties began, for a chlorine trifluoride motor operates at a chamber temperature close to 4000 K, where injectors and nozzle throats have a deplorable tendency to wash away, and unless the materials of which they are made are chosen with considerable astuteness, and unless the design is *very* good, the motor isn't going to last long. The propellant man liked CTF because of its performance, and the engineer hated the beast because it was so rough on motors and so miserable to handle. Although he had to learn to live with it, he postponed the learning process as long as he could. It is only recently, as the customers have been demanding a better performance than can be wrung out of IRFNA-UDMH, that CTF has been the subject of much intensive, large scale, testing.

Bromine pentafluoride, BrF_5 is very similar to ClF_3 as far as its handling properties are concerned, except that its boiling point (40.5°), is a little higher. Oddly enough, it never seems to perform as well as it should, and it's much harder to get a reasonable percentage of its theoretical performance out of it on the test stand than it is with CTF. Nobody knows why.

Very early in the game it was apparent to several of us in propellant chemistry that there really wasn't any fuel available that was right for ClF_3. Ammonia's performance was too low, and hydrazine, with an excellent performance and density, froze at a temperature that was much too high. And everything else had carbon in it. And with a fluorine type oxidizer that is bad. (See the chapter on performance.) It degrades the performance, and produces a conspicuous smoky exhaust stream. So in the latter part of 1958 Tom Reinhardt of Bell, Stan Tannenbaum of RMI and I at NARTS, unknown to each other, tried to do something about it. And since chemists with similar problems are likely to come up with similar answers, we went about it in very much the same way. Stan and Tom considered that the best place to start was with MMH, CH_6N_2, which was about as close to hydrazine as you could get, and then get enough oxygen into the system to burn the single carbon to CO. And they did this by mixing one mole of water with one of MMH, to get a mixture equivalent to COH_8N_2. When this was burned with CTF the carbon and oxygen went to CO and the hydrogens burned to HCl and HF. The performance was somewhat below that of hydrazine, since considerable energy was wasted in decomposing the water, but it was still better than that of ammonia. And they found that they could add considerable hydrazine (0.85 moles to one of MMH) to the mixture without raising the freezing point above −54°. Bell Aerosystems now calls the mixture BAF-1185.

I started with MMH, too. But I remembered all the work we had done with hydrazine nitrate, $N_2H_5NO_3$, and used that as my oxygen carrier, mixing one mole of it with three of MMH. And I found that I could add a mole or two of straight hydrazine to the mix without ruining my freezing point. I wanted to do performance calculations, to see how it would compare with hydrazine, and phoned Jack Gordon of RMI to get the heat of formation of MMH and hydrazine nitrate. He was (and is) a walking compendium of thermodynamic data, so I wasn't too surprised that he had the figures on the tip of his tongue. But my subconscious filed the fact for future reference.

Anyway, I did the performance calculations, and the results looked good—about 95 percent of the performance of straight hydrazine, and no freezing point troubles. So we made up a lot of the stuff and ran it through the wringer, characterizing it as well as we could, which was pretty well. We ran card-gap tests* on it, and found that it was quite shock insensitive, in spite of all that oxidizing salt in it. It seemed to be a reasonably good answer to the problem, so we code-named it "Hydrazoid N," and stuck it on the shelf for the engineers when they would need it.

Then, one day, I got a phone call from Stan Tannenbaum. "John, will you do some card gaps for me?" (RMI wasn't equipped to do them, and RMI and my outfit always had a comfortable, off-the-cuff, forget the paperwork and what the brass don't know won't hurt them, sort of relationship, so I wasn't surprised at the request.)

"Sure, Stan, no problem. What's the stuff you want me to fire?"

He hesitated a moment, and then, "It's proprietary information and I'm afraid I can't tell . . ."

"(-bleep-) you, Stan," I interrupted amiably. "If you think I'm going to tell my people to fire something without knowing what's in it you've got rocks in the head."

---

* The card-gap test is used to determine the shock sensitivity of a potentially explosive liquid. A 50-gram block of tetryl (high explosive) is detonated beneath a 40 cc sample of the liquid in question, contained in a 3″ length of 1″ iron pipe sealed at the bottom with a thin sheet of Teflon. If the liquid detonates, it punches a hole in the target plate, of ⅜″ boiler plate, sitting on top of it. The sensitivity of the liquid is measured by the number of "cards," discs of 0.01″ thick cellulose acetate, which must be stacked between the tetryl and the sample to keep the latter from going off. Zero cards means relatively insensitive, a hundred cards means that you'd better forget the whole business. As may be imagined, the test is somewhat noisy, and best done some distance from human habitation, or, at least, from humans who can make their complaints stick.

A longer pause. I suspect that my reaction wasn't unexpected. Then, "Well, it's a substituted hydrazine with some oxidizing material . . ."

"Don't tell me, Stan," I broke in. My subconscious had put all the pieces together. "Let me tell you. You've got three moles of MMH and one of hydrazine nitrate and—"

"Who told you?" he demanded incredulously.

God forgive me, but I couldn't resist the line. "Oh, my spies are everywhere," I replied airily. "And it doesn't go off at zero cards anyway." And I hung up.

But two minutes later I was on the phone again, talking to the people in the rocket branch in Washington, and informing them that RMI's MHF-1 and NARTS's Hydrazoid were the same thing, that Stan Tannenbaum and I had come up independently with the same answer at the same time and that nobody had swiped anything from anybody. The time to stop that sort of rumor is before it starts!

A few years later (in 1961), thinking that if hydrazine nitrate was good, hydrazine perchlorate ought to be better, I put together Hydrazoid P, which consisted of one mole of the latter, $N_2H_5ClO_4$, four of MMH, and four of straight hydrazine. It was definitely superior to Hydrazoid N, with a performance 98 percent of that of hydrazine itself, and a somewhat higher density. In putting it together, though, I remembered previous experience with hydrazine perchlorate, and figured out a way to use it without ever isolating the dry salt, which is a procedure, as you may remember, to be avoided. Instead, I added the correct amount of ammonium perchlorate (nice and safe and easy to handle) to the hydrazine, and blew out the displaced ammonia with a stream of nitrogen. Then I added the MMH, and I was in business. The mixture turned out to be somewhat corrosive to stainless steel at 71° (hydrazine perchlorate in *hydrazine* is a strong acid) but its behavior when it was spilled was what scared the engineers. If it caught fire as it lay on the ground, it would burn quietly for some time, and then, as the hydrazine perchlorate became more concentrated, it would detonate—violently. (Hydrazoid N, or any similar mixture, it turned out, would do the same thing.)

It seemed likely that if the burning rate of the mixture could be increased so much that the combustion would take place in the liquid and not in the vapor phase, the perchlorate would never have a chance to get concentrated, and the problem might be licked. I knew, of course, that certain metal oxides and ions catalyzed hydrazine decomposition, but I didn't want this to happen except under combustion conditions. The answer seemed to be to wrap the ion in a protective structure of some sort, which would be stripped off at combustion temperatures. So I told one of the gang to make the acetylacetonate complex of every metal ion he could find in the stockroom.

He came up with a dozen or so, and we tried them out. Some of them did nothing at all. Others started decomposing the Hydrazoid P as soon as they got into solution. But the nickel acetylacetonate did a beautiful job. It did nothing at all at room temperature or in storage. But half a percent or so speeded up Hydrazoid P combustion, either in the air, or when we burned the stuff under pressure as a mono-propellant, by orders of magnitude. But when we did fire tests in the open, the results weren't so good. An uncertainty factor had been introduced into Hydrazoid burning, and instead of detonating every time it did it about one time in three. So the engineers were still afraid of it.

A pity, too. For the nickel complex gave the fuel a peculiarly beautiful purple color, and somehow I'd always wanted a purple propellant!

Other fuels for ClF_3 have been developed, but they're generally rather similar to those I've described, with the carbon in them balanced out to CO by the addition of oxygen, somehow, to the mixture. On the whole, the problem can be considered to be pretty well under control. The detonation hazard after a spill is important on the test stand, but *not* with a prepackaged missile.

While the preliminary work with CTF was going on, and people were trying to come up with a good fuel for it, they were also looking very hard at the oxides of chlorine and their derivatives. Cl_2O_7, with an endothermic heat of formation of $+63.4$ kcal/mole, was one of the most powerful liquid oxidizers known in the early 50's, and preliminary calculations showed that it should give a remarkably high performance with any number of fuels. It had, however, one slight drawback—it would detonate violently at the slightest provocation or none at all. From first to last, at least five laboratories tried to domesticate it, with no success at all. The approach was to hunt for additives which would desensitize or stabilize it—Olin Mathieson, alone, tried some seventy—and was a dismal failure.

The closely related perchloric acid, at first, appeared to be a more promising candidate. Its heat of formation was exothermic, at least, and so the acid should show little tendency to decompose to the elements. However, 100 percent perchloric acid, like nitric acid, is not entirely what it seems. An equilibrium exists in the concentrated acid:

$$3HClO_4 \rightleftharpoons Cl_2O_7 + H_3OClO_4$$

so that there is always some of the very sensitive oxide present waiting to make trouble. And when it triggers the perchloric acid, the latter decomposes, not to the elements, but to chlorine, oxygen, *and* H_2O, with the release of enough energy to scare anybody to death. I had been ruminating on this fact, and had an idea. The structure of perchloric acid can be written

H—O—Cl=O. Now, if the HO group were replaced by an F, to give F—Cl=O, what could the stuff decompose to? Certainly there weren't any obvious products whose formation would release a lot of energy, and the compound ought to be reasonably stable. And it should be a real nice oxidizer.

So, one day in the spring of 1954, Tom Reinhardt, then the chief engineer of NARTS, Dr. John Gall, director of research of Pennsalt Chemicals, and I were sitting around the table in my laboratory shooting the breeze and discussing propellants in general. John was trying to sell us NF$_3$, but we weren't interested in anything with a boiling point of −129°. Then I brought up the subject of this hypothetical derivative of perchloric acid, added my guess that it would probably be low boiling, but not so low that it couldn't be kept as a pressurized liquid at room temperature, and my further guess that it should be rather inert chemically "because of that hard shell of electrons around it." And then I asked, "John, can you make it for me?"

His reply, delivered with considerable self-satisfaction, was enough to break up the meeting—and start a new one. "It *has* been made, its properties are as you predicted, and, just by coincidence, we just hired the man who discovered it."

My delighted whoop woke up the firehouse dog half a mile away—and was the beginning of the perchloryl fluoride program. It seems that in 1951, some workers in Germany had treated sodium chlorate, NaClO$_3$, with fluorine gas and had obtained sodium fluoride and various unidentified gaseous products which they did not identify—but one of them, in hindsight, must have been perchloryl fluoride. Then, in 1952, Englebrecht and Atzwanger, in Austria, dissolved sodium perchlorate in anhydrous hydrofluoric acid, HF, and electrolyzed the solution, mainly, I suspect, to see what would happen. They collected the gases involved, sorted them out, and isolated perchloryl fluoride among them. Since hydrogen, fluorine, and a few other items were all mixed together, they were plagued by explosions, but managed to survive the process somehow. (Englebrecht was just naturally venturesome to the point of lunacy. One of his other exploits was the development of a fearsome cutting torch burning powdered aluminum with gaseous fluorine. He could slice through a concrete block with it, to the accompaniment of an horrendous display of sparks, flames, and fumes which suggested an inadequately controlled catastrophe.) I had missed the report of the discovery (it appeared in an Austrian journal which I didn't normally see) but Pennsalt apparently had not, and decided that Englebrecht was just the sort of person they wanted on their staff.

In June BuAer authorized NARTS to investigate perchloryl fluoride, and Pennsalt sent us thirty-three grams of it in October—painfully produced by Englebrecht's method. And then, while we tried to characterize the material, they started looking for a simpler way to make it. Dr. Barth-Wehrenalp of their laboratories came up with—and patented—a synthesis by which it could be made rather easily and cheaply. It worked by the reaction KClO$_4$ + (excess)

$HSO_3F \longrightarrow KHSO_4 + FClO_3$, which looks simpler than it is. Nobody really understands the reaction mechanism.

While we were characterizing it, Pennsalt was doing the same, and passing their results over to us, and in a few months we knew just about everything we wanted to know about it. It was a pleasure working with that outfit. I'd phone one day asking, say, for the viscosity as a function of temperature, and within a week they would have made the measurements (and measuring the viscosity of a liquid under its own vapor pressure isn't exactly easy) and passed the results on to me.

In 1955 we were ready for motor work, and Pennsalt shipped—or rather hand-carried—ten pounds of perchloryl fluoride to us. (It was made by the old process, since the new one wasn't yet ready, and cost us $540 per pound. We didn't mind. We'd expected it to cost a thousand!)

With ten pounds of it we were able to make small motor tests (the fuel was MMH) and found that we had a very fine oxidizer on our hands. Its performance with MMH was very close to that of ClF_3 with hydrazine, and there wasn't any freezing point trouble to worry about. It was hypergolic with MMH, but starts were hard[*], so we used a starting slug of RFNA. Later, Barth-Wehrenalp tried mixing a small amount of chloryl fluoride, ClO_2F with it, and got hypergolic ignition that way.[**] But what made the rocket mechanics happy—was the fact that it varied from all other oxidizers in that you just couldn't hurt yourself with it, unless, as Englebrecht suggested, "you drop a cylinder of it on your foot." Its toxicity was surprisingly low, it didn't attack either inflammables or human hide, it wouldn't set fire to you—in fact, it was a joy to live with.

What did it in, finally, was the fact that its density at room temperature was rather low, 1.411 compared to 1.809 for CTF, and since its critical temperature was only 95°, it had a very high coefficient of expansion. Its volume would increase by 20 percent between 25° and 71°, so your tanks always had to be oversized. It is, however, completely miscible with all-halogen oxidizers such as CTF, and can be added to the latter to help them burn carbon-containing fuels, which need oxygen. This will probably be its future role.

While PF (so called for security and in deference to the engineers, who were apparently quite incapable of pronouncing the word "perchloryl") was being investigated, the next candidate was about to make its appearance.

---

[*] It seems that liquid perchloryl fluoride reacts with liquid amines, hydrazines, or ammonia, $FClO_3 + H_2N$—R HF + O_3Cl—NH—R and the perchloramide-type compound is remarkably and violently explosive. Hence the hard starts.

[**] Chloryl fluoride, ClO_2F, was first reported by Schmitz and Schumacher in 1942. It is indecently reactive, and the hardest to keep of all the Cl-O-F compounds, since it apparently dissolves the protective metal fluoride coatings that make the storage of ClF_3 comparatively simple.

Several laboratories, at this time, were trying to come up with storable oxidizers with a better performance than ClF_3, and in 1957, Colburn and Kennedy, at Rohm and Haas, reacted nitrogen trifluoride* with copper turnings at 450° and produced N_2F_4 by the reaction $2NF_3 + Cu \longrightarrow CuF_2 + N_2F_4$.

Here was something interesting, and the propellant community leaped into the act with glad cries and both feet. Research went off in two directions—improving the synthetic method of hydrazine tetrafluoride, as it was called,** for one, and determining its physical properties and its chemistry for another.

Rohm and Haas came up with a somewhat esoteric, not to say peculiar, synthesis when they reacted NF_3 with hot arsenic, of all things. Stauffer Chemical reacted NF_3 with hot fluidized carbon in a reaction which was easy to control, but which gave a product grossly contaminated with large amounts of C_2F_6, just about impossible to remove. Du Pont developed a completely different synthesis, in which NF_3 and NO are reacted at 600° in a nickel flow tube to form N_2F_4 and NOF. Other syntheses took a route through difluoramine, HNF_2, which was made by reacting urea in aqueous solution with gaseous fluorine to form F_2NCONH_2, and then hydrolyzing this with hot sulfuric acid to liberate the HNF_2. The final step was to oxidize the difluoramine to N_2F_4. Callery Chemical Co. did this with sodium hypochlorite in a strongly alkaline solution; Aerojet, as well as Rohm and Haas, did it with ferric ion in acid solution. The Du Pont process, and the HNF_2–route syntheses are those used today.

(There was some desire to use HNF_2 itself as an oxidizer—its boiling point is $-23.6°$ and its density is greater than 1.4—but it is so violently explosive that the idea never got very far. When it is used as an intermediate, the drill is to make it as a gas and use it up immediately.)

Dinitrogentetrafluoride was definitely a high-energy oxidizer, with a high theoretical performance with fuels such as hydrazine. (Marantz and his group at NBS soon determined its heat of formation so that accurate calculations could be made) and when Aerojet, in 1962, burned it with hydrazine and with pentaborane they measured 95 to 98 percent of the theoretical performance.

---

* Making NF_3 is tricky enough. It's done by electrolyzing molten ammonium bifluoride, using graphite electrodes. They have to be graphite—if you use nickel you don't get any NF_3—and the yield depends upon who manufactured the graphite. Don't ask me why.

** N_2F_4 is an inorganic compound, and should have been named according to the nomenclature rules of inorganic chemistry, "dinitrogen tetrafluoride" in strict analogy to "dinitrogen tetroxide" for N_2O_4. Instead it was named by the nomenclature rules of organic chemistry, as a derivative of hydrazine. This sort of thing was happening all the time, as organic chemists tried to name inorganic compounds, and inorganic chemists made a mess of naming organics.

And it had a fairly good density—1.397 at its boiling point. But that boiling point was −73°, which put it out of the class of storable propellants.

And this fact led to the concept of "space-storable propellants." As you will remember, 1957 was the year of Sputnik 1, when the public suddenly realized that there might be something to this science fiction foolishness of space travel after all. Anything remotely connected with space had suddenly become eminently salable, and if the services weren't able to use N_2F_4 in missiles, perhaps the space agencies (NACA, later NASA) could use it in space. After all, the hard vacuum of space is a pretty good insulator, and when you have, in effect, a Dewar flask the size of the universe available, you can store a low-boiling liquid a long time. An arbitrary upper limit (−150°) was set for the boiling point of a space-storable, but the custom is to stretch this limit to include the propellant you want to sell. OF_2, boiling at −144.8° is considered a space storable, but if you want to call its ideal partner, methane, CH_4, boiling at −161.5° one too, nobody is going to complain too loudly.

NF_3 is a rather inert material, and its chemistry isn't too complicated, but N_2F_4 turned out to be a horse of another color, with a peculiarly rich and interesting chemistry. The propellant men were not exactly overjoyed by this development, since they much prefer to deal with an unenterprising propellant, which just sits in its tank, doing nothing, until they get around to burning it.

N_2F_4 reacts with water to form HF and various nitrogen oxides, with nitric oxide to form the unstable and brilliantly colored (purple) F_2NNO, and with a bewildering number of oxygen-containing compounds to form NF_3, NOF, N_2 and assorted nitrogen oxides, by reactions which are generally strongly dependent upon the exact conditions, and frequently affected by traces of water or nitrogen oxides, by the material of the reactor, and by everything else that the experimenter can (or cannot) think of. Many of its reactions result from the fact that it is always partially dissociated to $2NF_2$, just as N_2O_4 is always partially dissociated to $2NO_2$, and that the extent of the dissociation increases with the temperature. This is the way a halogen, such as Cl_2, behaves, and N_2F_4 can be considered to be a pseudohalogen. Niederhauser, at Rohm and Haas, thought that as such, it should add across a double bond, and reacted it, in the vapor phase, with ethylene—and came up with $F_2NCH_2CH_2NF_2$. The reaction proved to be general, and it led to many things, some of which will be described in the chapter on monopropellants.

---

* This boiling point was a surprise to many, who had expected that it would be somewhere near that of hydrazine, or around 100°. But some of us had noted that the boiling point of NF_3 was very near that of CF_4, and hence expected that of N_2F_4 to be not too far from that of C_2F_6, which is −79°. So some of us, at least, weren't disappointed, since we hadn't hoped for much.

The handling and characteristics of N_2F_4 are fairly well understood now, and it is undeniably a high-performing oxidizer, but it is difficult to assess its future role as a propellant. It's not going to be used for any military application, and liquid oxygen is better, and cheaper, in the big boosters. It *may* find some use, eventually, in deep space missions. A Saturn orbiter would have to coast for years before the burn which puts it in orbit, and even with the thermal insulation provided by empty space liquid oxygen might be hard to keep for that long. And N_2O_4 would probably be frozen solid.

When Kennedy and Colburn found dinitrogentetrafluoride they knew what they were hunting for. But the next oxidizer was discovered by people who were looking for something else.

It seems that at the beginning of 1960, Dr. Emil Lawton of Rocketdyne, armed with an Air Force contract, had an idea that looked wonderful at the time. It was to react chlorine trifluoride with difluoroamine,

$$ClF_3 + 3HNF_2 \longrightarrow 3HF + Cl(NF_2)_3$$

thusly. He put Dr. Donald Pilipovich, "Flip," on the job. Flip built himself a metal vacuum line and started in. But he didn't get what he wanted. He got mainly $ClNF_2$, plus a small quantity of "Compound X." Compound X showed a strong NF_2O^+ peak on the mass spectrometer, and the question was the source of the oxygen. He investigated, and found that the chlorine trifluoride he was using was heavily contaminated with $FClO_2$ and ClO_2.

Meanwhile, Dr. Walter Maya, of the same group, was making O_2F_2 by an electrical discharge in a mixture of fluorine and oxygen. And he got some air in his line, by accident, and came up with Compound X too.

Flip was tied up with another job at that time, so Maya took over the Compound X problem. He found that an electrical discharge in a mixture of air and fluorine would give X, but that a discharge in a mixture of oxygen and NF_3 did even better. Dr. Bartholomew Tuffly of their analytical group invented a gelled fluorocarbon gas chromatograph column to separate the X from the NF_3, and its mass spectrum and molecular weight identified it unambiguously as ONF_3 or the long-sought F_2NOF.

In the meantime a group at Allied Chemical, Drs. W. B. Fox, J. S. Mackenzie, and N. Vandercook, had been investigating the electrical discharge reaction of OF_2 with NF_3, and had taken the IR spectrum of an impure mixture around the middle of 1959, but had not identified their products. The two groups compared their results and spectra around January 1961, and found that they had the same compound. Nuclear magnetic resonance (NMR) spectroscopy showed that it was ONF_3, and not F_2NOF.

And the moral of this story is that it's always worth trying an electrical discharge on your mixtures when you're hunting for new compounds. You *never* know what will happen. Almost anything can.

Bill Fox's group soon found that ONF_3 could be synthesized by the photochemical fluorination of ONF, and by the flame fluorination of NO, with a fast quench. The last synthesis is best for relatively large scale production.

A little later, I was chairing a session on propellant synthesis at one of the big meetings, and found, on the program, that both Rocketdyne and Allied were reporting on ONF_3. I knew that they differed widely in their interpretations of the chemical bonding in the compound, so I rearranged the program to put the two papers back to back, in the hope of starting a fight. No luck, though—they were both too polite. Too bad.

Another meeting, some years later, had more interesting results. In June 1966, a symposium on fluorine chemistry was held at Ann Arbor and one of the papers, by Professor Neil Bartlett of the University of British Columbia, was to be on the discovery and properties of ONF_3. Bartlett, a virtuoso of fluorine chemistry, the discoverer of OIF_5 and of the xenon fluorides, had, of course, never heard of Rocketdyne's and Allied's classified research. But Bill Fox, seeing an advance program, hurriedly had his report on the compound declassified, and presented it immediately after Bartlett's, describing several methods of synthesis, and just about every interesting property of the compound. Bill did his best not to make Bartlett look foolish, and Bartlett grinned and shrugged it off—"well, back to the old vacuum rack"—but the incident is something that should be noted by the ivory tower types who are convinced of the intellectual (and moral) superiority of "pure" undirected research to the applied and directed sort.

The compound has been called nitrogen oxidetrifluoride, nitrosyl trifluoride, and trifluoroamine oxide. The first is probably preferable. It boils at $-87.5°$, and its density at that temperature is 1.547. It is much less active chemically than dinitrogentetrafluoride, and is hence much easier to handle. It is stable in most metals, reacts only very slowly with water or alkalis, or with glass or quartz even at $400°$. In these respects it is very similar to perchloryl fluoride, which has a similar compact and symmetrical tetrahedral structure, with no reactive electrons. It reacts with fluorinated olefins to form $C-O-NF_2$ structures, and with SbF_5 to form the interesting salt $ONF_2^+ SbF_6^-$.

Its potential as an oxidizer seems to be similar to that of N_2F_4, and it should be useful in deep space missions.

Rocket motors designed to operate only in deep space are generally designed to have a comparatively low chamber pressure—150 psia or less—and it takes less energy to inject the propellants than would be the case with motors designed for sea-level use, whose chamber pressure is usually around 1000 psia. (In a few years it will probably be 2500!) And for the low injection pressure requirements of the deep space motors, some of the "space storables" seem peculiarly well suited. During the coast period, they could be kept below their normal boiling points. Then as the time for their use approached, a small energy source (a small electrical heating coil or the like) could be employed to heat them up to a temperature at which their vapor pressure would be well above the low chamber pressure of the motor, and could itself, be the injection pressure source, just as an aerosol spray is expelled by its own vapor pressure. Dinitrogentetrafluoride, nitrogen oxidetrifluoride, as well as the long known nitryl fluoride, FNO_2, seem to be particularly suitable for this sort of application. Aerojet, during 1963, did a great deal of work along these lines, with complete success.

It's a good idea, when choosing a pair of "space storables," to choose a fuel and an oxidizer that have a common liquid (temperature) range. If they are stored next to each other during a mission that lasts several months, their temperatures are going to get closer and closer together, no matter how good the insulation is. And if the temperature toward which the two converge is one at which one propellant is a solid and the other is a gas, there are going to be difficulties when it comes time for them to go to work. Likewise, if the self-pressurizing type of injection is used, design problems are simplified if the two have vapor pressures that are pretty close to each other. So, if the designer intends to use ONF_3, with a boiling point of $-87.5°$, ethane, whose boiling point is $-88.6°$, would be a good choice for the fuel.

Two space-storable systems have been investigated rather intensively. RMI and JPL, starting in 1963 or so, and continuing into 1969, worked out the diborane–OF_2 system, while Pratt and Whitney, Rocketdyne, and TRW, with NASA contracts, as well as NASA itself, have concentrated their efforts on OF_2 and the light hydrocarbons: methane, ethane, propane, 1-butene, and assorted mixtures of these. (In most of their motor work, they used a mixture of oxygen and fluorine as a reasonably inexpensive surrogate for OF_2.) All the hydrocarbons were good fuels, but methane was in a class by itself as a coolant, transpiration or regenerative, besides having the best performance. The OF_2–methane combination is an extremely promising one. (It took a long time for Winkler's fuel of 1930 to come into its own!)

The last part of the oxidizer story that I can tell without getting into trouble with Security is the saga of "Compound A." If I tell it in more detail than usual, the reasons are simple. The discovery of "A" is probably the most important achievement to date of the chemists who have made propellants a career, the story is well documented, and it illustrates admirably the nontechnical, but bureaucratic and personal obstacles they had to surmount.

While Walter Maya was doing electrical discharge experiments in 1960-61 (he made NF_3 that way, something that no one else had been able to do, and was trying to get things like N_3F_5) he occasionally got trace quantities of two compounds, with absorption bands at 13.7 and 14.3 microns, respectively, in the infra red. And for convenience he called them "Compound A" and "Compound B." At that point, he got tied up in another job, and Lawton put

Dr. Hans Bauer to the problem of identifying them. Bauer made slow progress, but finally got enough A to subject it to mass spectroscopy. And found that it had chlorine in it. Since only nitrogen and fluorine had been put into the apparatus, this took some explaining, and it seemed likely that the chlorotrifluorohydrocarbon (Kel-F) grease used on the stopcocks of the apparatus was entering into the reaction. Lawton had Bauer (much against his will) introduce some chlorine into the system, and it soon was obvious that only chlorine and fluorine were needed to make "A." From this fact, from the further fact that "A" reacted with traces of water to form $FClO_2$, and from the IR spectrum, Lawton suggested in a report submitted in September 1961, that "A" was ClF_5. At that precise moment Rocketdyne's contract (supported by the Advanced Research Projects Administration—ARPA—and monitored by the Office of Naval Research—ONR) was canceled.

It seems that somebody in Rocketdyne's solid propellant operation in Texas, several hundred miles away, had made a security goof regarding the ARPA program, and Dr. Jean Mock of ARPA felt that something had to be done by way of reproof. Besides, as he remarked to Dr. Bob Thompson, Lawton's boss, "Lawton claimed he made ClF_5 and we know that's impossible." So the project lay dormant for half a year.

Then, about March 1962, Dr. Thompson scraped up some company R and D money, and told Lawton that he'd support two chemists for three months, doing anything that Lawton wanted to do. Maya was put back on the job, and with Dave Sheehan's help, managed to make enough "A" to get an approximate molecular weight. It was 127—as compared with the calculated value of 130.5.

Armed with this information, Lawton went back to ARPA and pleaded with Dick Holtzman, Mock's lieutenant. Holtzman threw him out of the office. By this time it was the middle of 1962.

At this time Lawton had an Air Force research program, and he decided, in desperation, to use their program—and money—to try to solve the problem. The catch was that the AF program didn't allow for work on interhalogens, but apparently he figured that if he succeeded all would be forgiven. (In the old Royal Spanish Army there was a decoration awarded to a general who won a battle fought against orders. Of course, if he *lost* it, he was shot.) Pilipovitch was Lawton's Responsible Scientist by that time, and he put Dick Wilson on the job. And within a week he had come up with

$$ClF_3 + F_2 \longrightarrow ClF_5$$
$$ClF + 2F_2 \longrightarrow ClF_5$$
$$Cl_2 + 5F_2 \longrightarrow 2ClF_5$$
$$CsClF_4 + F_2 \longrightarrow CsF + ClF_5,$$

all four reactions requiring heat and pressure.

The next problem was to explain all this to the Air Force. It wasn't easy. When Rocketdyne's report got to Edwards Air Force Base in January 1963 the (bleep) hit the fan. Don McGregor, who had been monitoring Lawton's program, was utterly infuriated, and wanted to kill him—slowly. Forrest "Woody" Forbes wanted to give him a medal. There was a fabulous brouhaha, people were shifted around from one job to another, and it took weeks for things to settle down. Lawton was forgiven, Dick Holtzman apologized handsomely for ARPA and gave Lawton a new contract, and relative peace descended upon the propellant business. And when I heard, a few weeks later, of the discovery of ClF_5 (the code name, Compound "A" was kept for some years for security reasons) I sent Emil a letter which started, "Congratulations, you S.O.B.! I only wish I'd done it myself!" He was inordinately proud of it, and showed it to everybody at Rocketdyne.

ClF_5 is very similar to ClF_3, but, with a given fuel, has a performance about twenty seconds better. It boils at $-13.6°$, has a density of 1.735 at $25°$. And all of the techniques developed for using and handling CTF could be applied, unchanged, to the new oxidizer. To say that the propellant community was enthusiastic would be a mad under-statement.

On their ARPA contract the Rocketdyne group, by grace of Dick Wilson's tremendous laboratory skill, came up with "Florox"—but that one's still classified, and I can't talk about it without getting into trouble.* But nobody has yet come up with what $OClF_5$, which I called "Compound Omega," because it would be just about the ultimate possible storable oxidizer. It would be particularly useful with a fuel containing carbon, such as monomethyl hydrazine, CH_6N_2, with which it would react, mole for mole, to produce $5HF + HCl + CO + N_2$—a set of exhaust species to warm the heart of any thermodynamicist. Lawton and company tried, and are presumably still trying to get it, and Dr. Sam Hashman and Joe Smith, of my own group, hunted for it for more than three years, without any luck, although they employed every known synthetic technique short of sacrificing a virgin to the moon. (A critical shortage of raw material held that one up.) If anybody ever *does* synthesize Omega, it will probably be Neil Bartlett or somebody in Lawton's group.

A good deal of work has been done with mixed oxidizers, tailoring the mixture to match the intended fuel. NOTS for one, experimented in 1962 with "Triflox," a mixture of ClF_3, $FClO_3$ and N_2F_4, and Pennsalt, for another, examined "Halox," comprising ClF_3 and $FClO_3$. In this connection, it seems to me

---

* Emil Lawton has recently informed me (9/71) that Florox has been declassified since a Frenchman reported it independently late in 1970. It is $OClF_3$, and is made by the fluorination of Cl_2O or, of all things, chlorine nitrate, or $ClONO_2$. Its boiling point is $30.6°$, and it has a high density, 1.852. And since it contains oxygen, it can be used with a carbon-containing fuel, such as UDMH.

that a suitable mixture of ClF_5 and $FClO_3$ might be almost as good as the elusive Omega to burn with MMH.

One attempt to upgrade the performance of ClF_5 by adding N_2F_4 to it came to an abrupt end when the vapor pressure of the liquid mixture (stored in steel pressure bottles) started to rise in an alarming manner. It seems that the two oxidizers reacted thus:

$$ClF_5 + N_2F_4 \longrightarrow ClF_3 + 2NF_3.$$

And there was absolutely nothing that could be done about it.

Oh, yes. About "Compound B." That's a sad story. It turned out to be tungsten hexafluoride—WF_6—apparently from the tungsten filament in the mass spectrometer. Even Lawton can't win 'em all!

7

Performance

Since I've been talking about "performance" for some thousands of well chosen (I hope) words, it might not be a bad idea for me to explain, at this point, exactly what I mean by the word.

The object of a rocket motor is to produce thrust—a force. This it does by ejecting a stream of gas at high velocity. And the thrust is dependent upon two factors, the rate at which the gas is being ejected, in, say, kilograms per second, and the *velocity* at which it is ejected. Multiply rate by velocity and you get thrust. Thus, kilograms per second times meters per second gives the thrust in Newtons. (That is, if you're a man of sense and are working with the MKS Systéme Internationale of units.) If you want to increase your thrust you can do it either by increasing the mass flow (building a bigger motor) or by increasing the jet velocity, which generally means looking for a better propellant combination. The performance of a propellant combination is simply the jet velocity it produces.

Sometimes people not in the rocket business ask what is the "power" of, say, the Saturn V rocket. Power isn't a very useful concept in rocketry, since what you're trying to give your vehicle is momentum, which is proportional to the thrust times the time it is exerted. But if you define the power as the rate at which thermal or chemical energy is being converted to kinetic energy in the exhaust stream, a meaningful figure can be dug out. The kinetic energy of a given mass of exhaust gas (relative to the rocket, that is, not to the Earth or the Moon or Mars) is $Mc^2/2$, where M is the mass, and c is the velocity (again, relative to the rocket). And the power, or rate of energy conversion, is $\dot{M}c^2/2$, where $\dot{M}$ is the mass flow—kilograms per second, say. But, as we saw above, $\dot{M}c = F$, the thrust. So, putting these together, Power = $Fc/2$. Nothing simpler. Let us now proceed to Saturn V.

Saturn V has a thrust of 7,500,000 pounds *force.* Not mass, mind you; the distinction is important. That is equal to 33.36×10^6 Newtons. (One pound force = 4.448 Newtons, the MKS unit of force. That's a nice thing about MKS—there's no confusion between mass and force!) I don't remember the exact exhaust velocity of the Saturn engines, but it can't be very far from 2500 meters per second. So, multiply 33.36×10^6 by 2.5×10^3 and divide by two—and out comes the power, neatly in watts.

And the power so calculated is

$$41.7 \times 10^9 \text{ Watts}$$
or
$$41.7 \times 10^6 \text{ Kilowatts}$$
or
$$41.7 \times 10^3 \text{ Megawatts,}$$

which amounts to some 56 *million* horsepower. For comparison, the nuclear powerplant of the *Enterprise,* the most powerful afloat, generates some 300,000 HP. And the mass flow of propellants into the engines and exhaust gases out of the nozzles is some fifteen tons a second. Considered as the through-put of a chemical reactor—which it is—the figure is impressive.

So far, everything has been simple. But now things begin to get a little sticky. For the question arises, "How do you calculate the exhaust velocity, c, that you can get out of a given pair of propellants, burned at a definite chamber pressure, and properly expanded through a nozzle?" As we saw above, the energy of a given mass of exhaust gas, $E = Mc^2/2$. Rearranging this, $c = (2E/M)^{1/2}$. As all of the propellant injected into a motor comes out as exhaust gas (we hope!), the "M" in that equation is also the mass of the propellant which produced the mass of the exhaust gas that we're considering. But the E is *not* equal to the thermal energy, H, in the exhaust gas before it was expanded. So, actually, $c = (2H/M \times \eta)^{1/2}$, where η is the efficiency of conversion of thermal to kinetic energy. And η depends upon the chamber pressure, upon the exhaust pressure, and upon the nature of the exhaust gas, both as it exists unexpanded in the chamber and as it changes during expansion.

So, obviously, we have to know the chemical composition of the gas in the chamber. That's the first step. And you can't take it by using simple stoichiometry. If you put two moles of hydrogen and one of oxygen into the chamber you do *not* come out with two of water. You will have H_2O there, of course. But you will also, because of the high temperature, have a lot of dissociation, and the other species present will be H, H_2, O, O_2, and OH. Six species in all, and you can't know, a priori, in what proportions they will appear. And to solve for six unknowns you need six equations.

Two of these are simple. The first is derived from the atomic ratio between hydrogen and oxygen, and simply states that the sum of the partial pressures of all the hydrogen-bearing species, each multiplied by the number of hydrogen atoms in it, all divided by the sum of the partial pressures of all of the oxygen bearing species, each multiplied by the number of oxygen atoms in *it,* is a certain value upon which you have already decided, in this case two. The second equation states that the sum of the partial pressures of all the species present shall equal the chamber pressure which you have chosen. The other four equations are equilibrium equations of the type $(H)^2/(H_2) = K_1$ where (H) and (H_2) represent the partial pressures of those species, and K_1 is the constant for the equilibrium between them at the chamber temperature. This is a very simple case. It gets worse exponentially as the number of different elements and the number of possible species increases. With a system containing carbon, hydrogen, oxygen, and nitrogen, you may have to consider fifteen species or more. And if you toss in some boron, say, or aluminum, and perhaps a little chlorine and fluorine—the mind boggles.

But you're stuck with it (remember, I didn't *ask* you to do this!) and proceed—or did in the unhappy days before computers. First, you make a guess at the chamber temperature. (Experience helps a lot here!) You then look up the relevant equilibrium constants for your chosen temperature. Devoted and masochistic savants have spent years in determining and compiling these. Your equations are now before you, waiting to be solved. It is rarely possible to do this directly. So you guess at the partial pressures of what you think will be the major constituents of the mixture (again, experience is a great help) and calculate the others from them. You add them all up, and see if they agree with the predetermined chamber pressure. They don't, of course, so you go back and readjust your first guess, and try again. And again. And eventually all your species are in equilibrium and you have the right ratio of hydrogen to oxygen and so on, and they add up to the right chamber pressure.

Next, you calculate the amount of heat which would have been evolved in the formation of these species from your propellants, and compare that figure with the heat that would be needed to warm the combustion products up to your chosen chamber temperature. (The same devoted savants have included the necessary heats of formation and heat capacities in their compilations.) And, of course, the two figures disagree, so you're back on square one to guess another chamber temperature. And so on.

But all things come to an end, and eventually your heat (enthalpy) all balances, your equilibria all agree, your chamber pressure adds up, and you have the right elemental ratios. In short, you know the chamber conditions.

The next morning (the procedure described above has probably taken all day) you have to make a decision. Shall you make a frozen equilibrium calculation, or shall you make a shifting equilibrium calculation? If the first, you assume that the composition of the gas and its heat capacity remains unchanged as it is expanded and cooled in the nozzle. If the latter, you assume that as the gases cool and expand the equilibria among the species shift in accordance with the changing pressure and temperature, so that neither the composition nor the heat capacity of the exhaust gas is identical with what it was in the chamber. The first assumption amounts to a statement that all reaction rates are zero, the second to a statement that they are infinite, and both assumptions are demonstrably false.

If you want a conservative figure, you choose to make a frozen equilibrium calculation. (It gives a lower value than a shifting equilibrium calculation.) And you plug the data from the chamber calculations into the following horrendous formula.

$$c = \left\{ 2 \frac{R\gamma}{\gamma - 1} \frac{Tc}{\bar{M}} \left[1 - \left(\frac{Pe}{Pc}\right)^{\frac{\gamma-1}{\gamma}} \right] \right\}^{1/2}$$

Here, R is the universal gas contant, γ is the ratio of specific heats, Cp/Cv of the chamber gases. $\bar{M}$ is their average molecular weight. Tc is the chamber temperature. Pe and Pc are the exhaust and chamber pressures respectively. This formula looks like a mess, and it is, but it can be simplified to

$$c = \left[2H/M \right]^{1/2} \left[1 - \left(\frac{Pe}{Pc}\right)^{R/Cp} \right]^{1/2}$$

where H is the sum of the enthalpies of all the species present. (The reference state of zero enthalpy is taken to be the perfect gas at absolute zero.) "M," of course is the mass of propellants which produced them. And the efficiency, η, is

$$1 - \left(\frac{Pe}{Pc}\right)^{R/Cp}$$

If you feel optimistic—and energetic—you make a shifting equilibrium calculation. This is based on the assumption that although the gas composition will change during the expansion process, the entropy will not. So your next step is to add up the entropies of all the species present in the chamber, and put the figure on a piece of paper where you won't forget it. (Entropies are in the compilations, too.) Then, you guess at the exhaust temperature, at the exhaust pressure you have decided upon. And then you determine the composition of the exhaust gas, just as you did the chamber composition. And add up the entropies, there, and compare it with the chamber entropy. And try another exhaust temperature, and so on. Finally you have the exhaust conditions, and can calculate the enthalpy per unit mass there. And then, finally,

$$c = \left[\frac{2(H_c - H_e)}{M} \right]^{1/2}, \eta = (H_c - H_e)/H_c.$$

Solid and liquid exhaust products complicate the process somewhat when they appear, but that's the general idea. There is nothing complicated about it, but the execution is insufferably tedious. And yet I know people who have been doing performance calculations for twenty years and are still apparently sane!

The time and labor involved in an "exact" performance calculation had two quite predictable consequences. The first was that those calculations which *were* made were cherished as fine gold (for shifting equilibrium calculations read "platinum"), circulated, compiled, and squirreled away by anyone who could get his hands on them. The second consequence was that everybody and his uncle was demanding an approximate, or short method. And these were forthcoming, in considerable variety.

The most elaborate of these took the form of Mollier charts of the combustion products of various propellant combinations. These usually plotted enthalpy versus entropy, with isotherms and isobars cutting across the chart. A typical set of charts would be for the combustion products of jet fuel with various proportions of oxygen. Another, the decomposition products of 90 percent peroxide, another, ammonia and oxygen, at various O/F ratios. Some were more general, applying to a defined mixture of carbon, oxygen, hydrogen, and nitrogen atoms, without specifying what propellants were involved. These charts were easy to use, and gave results in a hurry, but they seldom applied to *exactly* the combination you had in mind. They were also very difficult to construct, involving, as they did, dozens of calculations. The Bureau of Mines, with its extensive experience with combustion phenomena, was a leader in this field.

A more general but less informative method was developed in 1949 by Hottel, Satterfield, and Williams at MIT. This could be used for practically any combination in the CHON system, but using it for any chamber pressure other than 300 psia, or any exhaust pressure other than 14.7, was an involved and messy procedure. I later modified and streamlined the method, and made some provision for other elements, and published it in 1955 as the "NARTS Method of Performance Calculation."

These, and similar graphical methods, involve, essentially, an interpolation between accurately calculated systems, and they gave a fairly good approximation of the results of a shifting equilibrium calculation.

The other group of methods gave, generally, results that approximated those of a frozen equilibrium calculation, and were based on the equation $c = (2H/M \times \eta)^{1/2}$. The usual procedure was to determine H by ignoring any minor products (pretending that there wasn't any dissociation). The products in the CHON system were assumed to be CO_2, H_2O, CO, H_2, and N_2. Once the water-gas equilibrium was determined (that was done by using the equilibrium constant at some arbitrary temperature, such as 2000 K, or at the whim of the operator—it didn't matter too much)˙ H could be determined by simple arithmetic. As for η, with a little experience you could make a pretty good guess at it, and any error would be halved when you took the square root of your guess! Or, if you wanted to be fancy, you could determine the average C_p of your gases at somewhere near what you thought your chamber temperature ought to be, and plug that into the efficiency term. Tom Reinhardt's 1947 method included curves of temperature vs enthalpy for various exhaust gases, as well as C_p vs temperature. You determined your temperature from your enthalpy, and the C_p from the temperature. The temperature, of course, was much too high, since dissociation was ignored. Ten years later I modified the method, eliminating the curves, devising a fast and easy way of getting an R/C_p averaged over the whole temperature range, and providing a nomograph for calculating η from that and the pressure ratio. It was called the NQD—NARTS Quick and Dirty—method. The thing worked astoundingly well, giving results agreeing with complete shifting equilibrium calculations (I suppose that the averaged R/C_p helped there) to something like 1 percent. And you could make a calculation in fifteen minutes. It worked best, too, when you postulated the simplest—in fact the most simple minded—set of products imaginable. And it was adaptable. When a man from Callery Chemical Co. came in one day and told me for the first time about the BN system I learned that. In this system the exhaust products are hydrogen and solid BN. I hauled out my tables when he told me about it, and letting two atoms of carbon (graphite) pretend that they were one of molecule of BN, made a fast estimate. And lit on the nose. My value was within half a percent of the one he had obtained from a fancy machine calculation. The only trouble with the method was that I never could keep a copy for myself. Some character was always mooching my last copy, and I'd have to run off another fifty or so.

There were other approximate methods developed, some as late as 1963, but they were all similar to those I've described. But the day of the shorthand method is gone—as is, thank God!—the complete hand calculation.

The computers started getting into the act in the early 50's, although considerable chemical sophistication was needed to make the most of their initially somewhat limited capabilities. At Bell Aerosystems they were considering fluorine as an oxidizer, and a mixture of hydrazine and methanol as the fuel, and demanded performance calculations. The programmer protested that he couldn't handle that many elements, and Tom Reinhardt retorted, "The carbon and the oxygen will go to CO, and you just tell the little man who lives inside that box to treat it exactly like nitrogen," End of problem.

---

* Consider the case where one O_2, one H_2 and one C react. If the reaction went to $H_2O + CO$, the performance would vary by only 2.5 percent from the performance if it went to CO_2 and H_2. And this is the worst possible case!

All the compilations of thermodynamic data are on punch cards, now, versatile programs, which can handle a dozen or so elements, are on tape, and things are a lot simpler than they were. But the chemical sophistication is still useful, as is a little common sense in interpreting the print-out. As an example of the first, calculations were made for years on systems containing aluminum, using thermodynamic data on gaseous Al_2O_3 calculated from its assumed structure. And the results didn't agree too well with the experimental performances. And then an inconsiderate investigator proved that gaseous Al_2O_3 didn't exist. Red faces all over the place. As an example of the second, consider the case of a propellant combination that produces a lot of solid carbon, say, in the exhaust stream. The machine makes its calculations on the assumption that the carbon is in complete thermal and mechanical equilibrium with the gaseous part of the exhaust. A bit of common sense suggests that this will not be so, since heat transfer is not an infinitely rapid process, and that the carbon may well be exhausted considerably hotter than the surrounding gas. So you look at the print-out with considerable pessimism—and wait for experimental results before committing yourself. A great deal of effort, in recent years, has gone into attempts to develop programs which will take things like heat transfer from solid to gas into account, and which will allow for the actual velocity of the change in the exhaust composition during expansion. These are called "kinetic" programs, as opposed to the frozen or shifting equilibrium programs, and only the big computers make them possible. There is only one trouble with them. Reliable kinetic data are as hard to come by as honest aldermen—and when you feed questionable data into the machine, questionable results come out at the other end. As the computer boys say, "Garbage in—garbage out."

And there is one disconcerting thing about working with a computer—it's likely to talk back to you. You make some tiny mistake in your FORTRAN language—putting a letter in the wrong column, say, or omitting a comma—and the 360 comes to a screeching halt and prints out rude remarks, like "ILLEGAL FORMAT," or "UNKNOWN PROBLEM," or, if the man who wrote the program was really feeling nasty that morning, "WHAT'S THE MATTER STUPID? CAN'T YOU READ?" Everyone who uses a computer frequently has had, from time to time, a mad desire to attack the precocious abacus with an axe.

Rocket performance is not usually reported in terms of exhaust velocity, although the early workers wrote in those terms. Instead, it is reported as "specific impulse," which is the exhaust velocity divided by the standard acceleration of gravity, 9.8 meters or 32.2 feet per second^2. This practice gives figures of a convenient size in the range of 200 to 400 or so, but it has led to some rather tortuous, if not ludicrous definitions. The most common one is that specific impulse is the thrust divided by the *weight* flow of propellant, and it comes out in seconds. Putting the acceleration of gravity into the equation did that, but specifying the performance of a rocket, whose whole job is to get away from the earth, in terms of the acceleration of gravity on the surface of that planet, seems to me to be a parochial, not to say a silly procedure. (The Germans, during World War II, used an even sillier measure of performance, "specific propellant consumption," which was the reciprocal of specific impulse. This didn't even have the virtue of producing figures of a convenient size, but gave things like 0.00426 per second.)

Probably the best way of thinking of specific impulse is as a velocity expressed, not in meters or feet per second, but in units of 9.8 meters (or 32.2 feet) per second. That way you retain the concept of mass flow, which *is* relevant everywhere, and doesn't depend upon the local properties of one particular planet, and at the same time lets European and American engineers understand each other. When he hears $I_s = 250$, the European multiplies by 9.8 to get the exhaust velocity in meters per second, while the American does the same with 32.2 and comes out with feet per second. (*When* will the U.S. ever change over to MKS?!)

I've told you what performance is, and I've described the way you go about calculating it. But now comes the practical problem of picking a propellant combination which will give you a good one. Here it will be helpful to go back to the velocity equation, $c = [2H/M]^{1/2} [1-(Pe/Pc)^{R/C_p}]^{1/2}$ and to consider the H/M term and the efficiency term separately. Obviously, you want to make H/M as large as possible. And to do this, it is useful to consider the exhaust gases you hope to get. The energy contributed by a molecule of combustion products equals the heat of formation of that molecule from its elements at 25°C, plus its sensible heat above absolute zero (this is a very small item) *minus* the energy required to break down to their elements, at 25°C, the propellants which formed it. This last term is generally much smaller than the first—otherwise we wouldn't have useful propellants. And sometimes it is negative; when a mole of hydrazine breaks down to hydrogen and nitrogen we get some twelve kilocalories as a free bonus. But the important item is the heat of formation of the product molecule. That we want as big as possible. And, obviously, to maximize H/M, we must minimize M. So, to get a good energy term, we need an exhaust molecule with a high heat of formation and a low molecular weight.

So far so good. But now let's look at the efficiency term. Obviously, we want to get it as close to 1.0 as possible, which means that we want to beat $\left(\dfrac{Pe}{Pc}\right)^{R/C_p}$ down as far as we can. P_e/P_c is, of course smaller than one, so to do this we must raise the exponent R/C_p as high as we can. Which, of course, means that we want exhaust products with as low a C_p as we can find. And so we are hunting for exhaust products which have:

a A high heat of formation.
b A low molecular weight.
c A low C_p.

Alas, such paragons among exhaust products are hard to come by. Generally, if you have a good H/M term, the R/C_p lerm is bad. And vice versa. And if both are good, the chamber temperature can get uncomfortably high.

If we consider specific exhaust products, this is what we find: N_2 and solid C are practically useless as energy producers. HCl, H_2, and CO are fair*. CO_2 is good, while B_2O_3, HBO_2, OBF, BF_3, H_2O, and HF, as well as solid B_2O_3 and Al_2O_3, are excellent. When we consider the R/Cp term, the order is quite different. The diatomic gases, with an R/Cp above 0.2, are excellent. They include HF, H_2, CO, HCl, and N_2. (Of course a monatomic gas has an R/C_p of 0.4, but finding a chemical reaction which will produce large quantities of hot helium is out of the range of practical politics.) The triatomic gases, H_2O, OBF, and CO_2, with an R/C_p between 0.12 and 0.15 are fair. The tetratomic HBO_2 and BF_3, at about 0.1, are poor, and B_2O_3—well, perhaps it should be passed over in silence. As for the solids, C, Al_2O_3, and B_2O_3, their R/C_p is precisely zero, as would be the thermal efficiency if they were ever the sole exhaust products.

Faced with this situation, all the rocket man can do is hunt for a reasonable compromise. He would, if he could, choose pure hydrogen as his exhaust gas, since at any given temperature one gram of hydrogen has more heat energy in it than a gram of any other molecule around (one gram of H_2 at 1000 K has almost ten times the energy of one of HF at the same temperature), and its excellent R/C_p makes it possible to use a large fraction of that energy for propulsion. So hydrogen is the ideal working fluid, and you always try to get as much of it as possible into your mix. For it has to be a mix (in a chemical rocket, anyway) since you need an energy source of some sort to heat that hydrogen up to 1000 K or 3000 K or whatever. And the only available energy source is the combustion of some of the hydrogen. So you bring some oxygen or fluorine into the picture, to burn part of the hydrogen to H_2O or HF, bringing the temperature up to 3000 K or so, and your exhaust gas is the mixture of H_2O or HF with the excess hydrogen. When hydrogen is the fuel, it is always used in excess, and never burned completely to water or HF. If it were, the chamber temperature would be uncomfortably high, and the R/C_p of the

---

* The classification of hydrogen, as a fair contributor of energy even though it, naturally, has a zero heat of formation, is explained by the fact that the molecule is so light. At 25° it has a sensible heat, or heat content of 2.024 kilocalories per mole above absolute zero, and since the molecular weight is only 2.016, its H/M, even at room temperature, is 1.0 Kcal/gm.

mixture would be lowered and the performance would drop. Hydrogen is so light that a considerable excess of it won't harm the H/M term appreciably, and you get the maximum performance, generally, when you use only enough oxygen or fluorine to burn perhaps half of your fuel.

If you're burning a hydrocarbon with oxygen, or if you're working with the CHON system in general, you generally get the maximum performance from a mixture ratio which gives a 1.05 to 1.20 ratio of reducing to oxidizing valences in the chamber—that is, you work a little on the rich side of stoichiometric to get some CO and H_2 into the mixture and improve R/C_p. ("Rich" and "lean" in the rocket business mean exactly what they do in a carburetor.)

If you're using a halogen oxidizer with a storable fuel, the best results generally show up if your mixture ratio makes the number of fluorine atoms (plus chlorine atoms, if any) exactly equal the number of hydrogen atoms. If there is any carbon in the combination, it's a good idea to get enough oxygen into the system to burn it to CO, so you won't have any solid carbon in the exhaust. And if your energy-producing species is a solid or liquid at the exhaust temperature—BeO, Al_2O_3 are examples—the thing to do, of course, is to cram as much hydrogen as possible into the combination.

These are just a few of the things that the propellant chemist has to consider when he's looking for performance. And coming up with propellant combinations which will perform as the engineers want them to is what he's paid for. Inadequately.

This is how he goes about it: The engineering group have been given the job of designing the propulsion system of a new surface-to-air missile—a SAM. It is specified by the customer that it must work at any temperature likely to be encountered in military operations. The maximum dimensions are fixed, so that the missile will fit on existing launchers. It must be a packaged job, loaded at the factory, so that propellants won't have to be handled in the field. It must not leave a visible trail, which would make countermeasures easier. And, of course, it must have a much higher performance than the present system, which burns acid-UDMH. (The customer probably makes a dozen more demands, most of them impossible, but that will do for a starter.)

The engineers, in turn, before sitting down to their drawing boards, demand of the propellant chemist that he produce a combination that will make the missile do what the customer wants it to do. They also add some impossible demands of their own.

The chemist crawls into his hole to consider the matter. What he'd like to recommend is the hydrazine-chlorine pentafluoride (for historical reasons, ClF_5 is generally called "compound A") combination. It has the highest performance of any practical storable combination known (all the exhaust products are diatomic, and $^2/_3$ of them are HF), and it has a nice fat density, so you can stuff a lot of it into a small tank. But he remembers that all-weather constraint, and reminds himself that you can never tell where you might have to fight a war, and that the freezing point of hydrazine is somewhat incompatible with the climate of Baffin Land. So—the next best bet is, probably MHF-3, a 14–86 mixture of hydrazine and methyl hydrazine with the empirical formula $C_{0.81}H_{5.62}N_2$. Its freezing point is down to the magic $-54°$. (There are other possible fuels, but they may be somewhat dangerous, and he *knows* that MHF-3 is safe, and works.) But, with ClF_5, MHF-3 would leave a trail of black smoke leading right back to the launcher—definitely undesirable if the crew of the latter want to live to fire another round. Also, his professional soul (it's the only soul he has left after all these years in the business) is revolted by the thought of that free carbon and its effect on the R/C_p term and what it will do to his performance.

So he decides to spike his oxidizer with a bit of oxygen to take care of the carbon. Which means spiking it with an oxygen containing storable oxidizer. The only one of these which can live with compound A is perchloryl fluoride, "PF." So PF it will be.

He knows that when you have carbon and hydrogen in your system, along with oxygen and fluorine and chlorine, you generally get the best performance when the oxygen and carbon balance out to CO, and the hydrogen and the halogens balance to HF and HCl. So he doodles around a bit, and comes up with the equation:

$$C_{0.81}H_{5.62}N_2 + 0.27ClO_3F + 0.8467ClF_5$$
$$= 0.81CO + N_2 + 1.1167HCl + 4.5033HF$$

That looks good—lots of HF and hence a lot of energy. And there's nothing but diatomic gases in the exhaust, which means a good R/C_p, which means, in turn, that a gratifyingly large fraction of that energy will go into propulsion. To find out what that fraction will be, he packs up his notes and pays a call on the IBM 360. The results of the consultation are pleasing, so he converts his mole fractions into weight percentages, and calls on the engineers.

"Your fuel is MHF-3," he announces, "and your oxidizer is 80 percent 'A' and 20 PF. And your O/F is 2.18. And Muttonhead says –" "Who's Muttonhead?" "Muttonhead's the computer. He says that the performance, shifting, at $1000/14.7$ pounds is 306.6 seconds, and *I* say that if you can't wring out 290 on the test stand you're not half as good as you say you are. But watch your O/F. If you're lean the performance will drop off in a hurry, and if you go rich you'll smoke like crazy: The density is 1.39, and the chamber temperature is 4160 K. If you want it in Fahrenheit, convert it yourself!"

He then retreats hurriedly to his lair, pursued by the imprecations of the engineers, who, (a) complain that the density is too low, and, (b) that the chamber temperature is much too high and who ever heard of anybody operating that hot anyway? (c) demand that he do something about the toxicity of ClF_5. To which he replies that (a) he'd like a higher density himself, but that he's a chemist and not a theologian and that to change the properties of a compound you have to consult God about it; (b) to get high performance you need energy, and that means a high chamber temperature, and unless they're satisfied with RFNA and UDMH they'll have to live with it, and for (c) see the answer to (a).

And then, for the next six months or so he's kept busy telling them, in response to complaints:

"No, you can't use butyl rubber O-rings with the oxidizer! Do you want to blow your head off?"

"No, you can't use them with the fuel either. They'll go to pieces."

"No, you can't use copper fittings with the fuel!"

"Of course, your mixture ratio goes off if you put five gallons of the oxidizer in a fifty-gallon tank! Most of the PF is up in the ullage, and most of the A is down in the bottom of the tank. Use a smaller tank."

"No, there isn't any additive I can put in the oxidizer that will reduce the vapor pressure of the PF."

"And no, I can't repeal the first law of thermodynamics. You'll have to talk to Congress!"

And he dreams wistfully of climbing into a cold Martini—and wonders why he ever got into this business.

8

Lox and Flox
and Cryogenics
in General

While all this was going on, liquid oxygen was still very much in the picture. The sounding rocket Viking burned it with ethyl alcohol, as had the A-4, and so did several experimental vehicles of the early 50's, as well as the Redstone missile. Most of these, too, used the auxiliary power source of the A-4, hydrogen peroxide, to drive the feed pumps, and so on. The X-1, the first supersonic plane, was driven by an RMI Lox-alcohol rocket motor.

Other alcohols were tried as fuels to be used with oxygen—methanol by JPL as early as 1946, and isopropanol by North American early in 1951—but they weren't any particular improvement over ethanol. Neither was methylal, $CH_3OCH_2OCH_3$, which Winternitz, at RMI, was pressured into trying, much against his will (he knew it was a lot of foolishness) early in 1951. It seems that his boss had a friend who had a lot of methylal on hand, and if only some use for it could be found—? And at NARTS we did some studies for Princeton, using LOX and pure USP type drinking alcohol—not the denatured stuff. The only difference we could find was that it evaporated a lot faster than denatured alcohol when a sailor opened a drum to take a density reading. We had some very happy sailors while that program was going on.

But something more potent than alcohol was needed for the X-15 rocket-driven supersonic research plane. Hydrazine was the first choice, but it sometimes exploded when used for regenerative cooling, and in 1949, when the program was conceived, there wasn't enough of it around anyway. Bob Truax of the Navy, along with Winternitz of Reaction Motors, which was to develop the 50,000 pounds thrust motor, settled on ammonia as a reasonably satisfactory second best. The oxygen-ammonia combination had been fired by JPL, but RMI really worked it out in the early 50's. The great stability of the ammonia molecule made it a tough customer to burn and from the beginning they were plagued with rough running and combustion instability. All sorts of additives to the fuel were tried in the hope of alleviating the condition, among them methylamine and acetylene. Twenty-two percent of the latter gave smooth combustion, but was dangerously unstable, and the mixture wasn't used long. The combustion problems were eventually cured by improving the injector design, but it was a long and noisy process. At night, I could hear the motor being fired, ten miles away over two ranges of hills, and could tell how far the injector design had progressed, just by the way the thing sounded. Even when the motor, finally, was running the way it should, and the first of the series was ready to be shipped to the West Coast to be test-flown by Scott Crossfield, everybody had his fingers crossed. Lou Rapp, of RMI, flying across the continent, found himself with a knowledgeable seat mate, obviously in the aerospace business, who asked him his opinion of the motor. Lou blew up, and declared, with gestures, that it was a mechanical monster, an accident looking for a place to happen, and that he, personally, considered that flying with it was merely a somewhat expensive method of suicide. Then, remembering something he turned to his companion and asked. "By the way, I didn't get your name. What is it?"

The reply was simple. "Oh, I'm Scott Crossfield."

Our first real IRBM's were Thor and Jupiter, and these were designed to burn oxygen and JP-4. And the pumps would be driven by a gas generator burning the same propellants, but with a very rich mixture, to produce gases which wouldn't melt the turbine blades. JP had a better performance than alcohol, and getting rid of the peroxide simplified matters.

But there were troubles. The sloppy specifications for JP-4 arose to haunt the engineers. It burned all right, and gave the performance it should—but. In the cooling passages it had a tendency to polymerize (you will remember that the specifications allowed a high percentage of olefins) into tarry substances which slowed the fuel flow, whereupon the motor would cleverly burn itself up. And in the gas generator it produced soot, coke, and other assorted deposits that completely fouled up the works. And, of course, no two barrels of it were alike. (Also, believe it or not, it grows bacteria which produce sludge!)

But they needed the performance of a hydrocarbon; alcohol would not do. So then what?

Finally somebody in authority sat down and thought the problem through. The specifications of JP-4 were as sloppy as they were to insure a large supply of the stuff under all circumstances. But Jupiter and Thor were designed and carry nuclear warheads, and it dawned upon the thinker that you a large and continuing supply of fuel for an arsenal of such missiles. e is fired, if at all, just once, and after a few dozen of them have been lobbed over by the contending parties, the problem of fuel for later salvos becomes academic, because everybody interested is dead. So the only consideration is that the missile works right the first time—and you can make your fuel specifications just as tight as you like. Your first load of fuel is the only one you'll ever need.

The result was the specification for RP-1, which was issued in January of 1957. The freezing point limit was −40°, the maximum olefin content was set at 1 percent, and of aromatics at 5 percent. As delivered, it's usually better than the specifications: a kerosene in the C_{12} region, with a H/C ratio between 1.95 and 2.00, containing about 41 percent normal and branched paraffins, 56 of naphthenes, three of aromatics, and no olefins at all.

The polymerization and coking problems were solved, but Madoff and Silverman, at Rocketdyne (which was the autonomous division formed at North American to do all their rocket work) weren't entirely happy with the solution, and did extensive experimentation with diethylcyclohexane which, while not a pure compound, was a highly reproducible mixture of isomers, and was easy to come by. The results of their experiments were excellent, the fuel being appreciably superior to RP-1, but it never got into an operational missile. Atlas and Titan I, our first ICBM's were designed around RP-1 before Madoff and Silverman did their work, and Titan II used storable propellants. The F-1 motors of Saturn V burn LOX and RP-1.*

Oxygen motors generally run hot, and heat transfer to the walls is at a fantastic rate. This had been a problem from the beginning, even with regenerative cooling, but in the spring of 1948 experimenters at General Electric came up with an ingenious fix. They put 10 percent of ethyl silicate in their fuel, which was, in this case, methanol. The silicate had the happy faculty of decomposing at the hot spots and depositing a layer of silicon dioxide, which acted as insulation and cut down the heat flux. And, although it was continuously ablated and swept away, it was continuously redeposited. Three years later, also at GE, Mullaney put 1 percent of GE silicone oil in isopropanol, and reduced the heat flux by 45 percent. The GE first stage motor of Vanguard used such a heat barrier. Winternitz at RMI had similar good results in 1950 and 1951 with

* LOX and RP-1 never burn absolutely clean, and there is always a bit of free carbon in the exhaust, which produces a luminous flame. So when you're looking at TV and see a liftoff from Cape Kennedy—or from Baikonur for that matter—and the exhaust flame is very bright, you can be sure that the propellants are Lox and RP-1 or the equivalent. If the flame is nearly invisible, and you can see the shock diamonds in the exhaust, you're probably watching a Titan II booster burning N_2O_4 and 50–50.

ethyl silicate in ethanol and in methylal, and in 1951, with 5 percent of it in ammonia, he cut the heat flux by 60 percent.

Another tricky problem with an oxygen motor is that of getting it started. From the A-4 to Thor and Jupiter, a pyrotechnic start was the usual thing, but the complications were considerable and the reliability was poor. Sänger had used a starting slug of diethyl zinc, and Bell Aerosystems, in 1957, went him one better by using one of triethyl aluminum to start an oxygen-JP-4 motor. This technique was used in the later Atlas and all subsequent oxygen-RP motors. A sealed ampoule containing a mixture of 15 percent triethyl aluminum and 85 percent of triethyl boron is ruptured by the pressure in the fuel lines at start-up, reacts hypergolically with the liquid oxygen, and you're in business. Simple, and very reliable.

Alcohol, ammonia, and JP-4 or RP-1 were the fuels usually burned with LOX, but practically every other inflammable liquid available has been tried experimentally at one time or another. RMI tried, for instance, cyclopropane, ethylene, methyl acetylene, and methyl amine. None of these was any particular improvement on the usual fuels. Hydrazine was tried as early as 1947 (by the Bureau of Aeronautics at EES, Annapolis) and UDMH was tried by Aerojet in 1954. But in this country, in contrast to Russia, the combination of a hydrazine fuel and liquid oxygen is unusual. The only large-scale use of it was in the Jupiter-C, and the Juno-1 which were propelled by uprated Redstone motors, redesigned to burn Hydyne rather than alcohol. (Hydyne is a Rocketdyne developed 60–40 mixture of UDMH and diethylene triamine.)

Tsiolkovsky's ideal fuel was, of course, liquid hydrogen. It is useless, naturally, in a missile (its density is so low that it takes an inordinate tankage volume to hold any great amount of it) and the engineering problems stemming from its low boiling point are formidable, so it was pretty well left alone until after World War II.

Even then, it wasn't exactly easy to come by. There were just three organizations equipped to produce liquid hydrogen in 1947: the University of Chicago, the University of California, and Ohio State, and their combined productive capacity was 85 liters, or 13 pounds, per hour. (Assuming that the equipment could be run continuously, which it could not.) But in 1948 H. L. Johnson, of the Ohio State Research Foundation, burned it with oxygen in a small motor of about 100 pounds thrust. The next year Aerojet installed a 90-liter per hour continuous unit, and raised the U.S. capacity to 27 pounds an hour. Aerojet fired it at the 3000-pound thrust level, and used it as a regenerative coolant. (Each of the six 200,000 pound hydrogen motors in Saturn V, five in the second stage, one in the third, burns 80 pounds of hydrogen per *second*.)

Hydrogen is a super-cryogenic. Its boiling point of 21 K is lower than that of any other substance in the universe except helium. (That of oxygen is 90 K.)

Which means that problems of thermal insulation are infinitely more difficult than with oxygen. And there is another difficulty, which is unique to hydrogen.

Quantum mechanics had predicted that the hydrogen molecule, H_2, should appear in two forms: ortho, with the nucleii of the two atoms spinning in the same direction (parallel), and para, with the two nucleii spinning in opposite directions (antiparallel). It further predicted that at room temperature or above, three-quarters of the molecules in a mass of hydrogen should appear in the ortho form and a quarter in the para, and that at its boiling point almost all of them should appear in the para state.

But for years nobody observed this phenomenon. (The two forms should be distinguishable by their thermal conductivity.) Then, in 1927, D. M. Dennison pointed out, in the *Proceedings of the Royal Society,* that the transition from the ortho to the para state might be a slow process, taking, perhaps, several days, and that if the investigators waited a while before making their measurements, they might get some interesting results.

Urey, Brickwedde and others in this country, as well as Clusius and Hiller in Germany looked into the question exhaustively between 1929 and 1937, and the results were indeed interesting, and when the propellant community got around to looking them up, disconcerting. The transition *was* slow, and took several days at 21 K. But that didn't matter to the rocket man who merely wanted to burn the stuff. What did matter was that each mole of hydrogen (2 grams) which changed from the ortho to the para state gave off 337 calories of heat in the process. And since it takes only 219 calories to vaporize one mole of hydrogen, you were in real trouble. For if you liquefied a mass of hydrogen, getting a liquid that was still almost three quarters orthohydrogen, the heat of the subsequent transition of that to parahydrogen was enough to change the whole lot right back to the gaseous state. All without the help of any heat leaking in from the outside.

The answer to the problem was obvious—find a catalyst that will speed up the transition, so that the evolved heat can be disposed of during the cooling and liquefaction process and won't appear later to give you trouble; and through the 50's, several men were looking for such a thing. P. L. Barrick, working at the University of Colorado and at the Bureau of Standards at Boulder, Colorado, came up with the first one to be used on a large scale—hydrated ferric oxide. Since then several other catalytic materials have been found—palladium–silver alloys, ruthenium, and what not, several of them much more efficient than the ferric oxide—and the ortho-para problem can be filed and forgotten.

By 1961 liquid hydrogen was a commercial product, with Linde, Air Products, and several other organizations ready to sell you any amount you wanted, and to ship it to you in tank car lots. (The design of those tank cars, by the way, is quite something. Entirely new *kinds* of insulation had to be invented to make them possible.)

Handling liquid hydrogen, then, has become a routine job, although it has to be treated with respect. If it gets loose, of course, it's a ferocious fire and explosion hazard, and all sorts of precautions have to be taken to make sure that oxygen doesn't get into the stuff, freeze, and produce a murderously touchy explosive. And there is a delightful extra something about a hydrogen fire—the flame is almost invisible, and at least in daylight, you can easily walk right into one without seeing it.

A rather interesting recent development is slurried, or "slush" hydrogen. This is liquid hydrogen which has been cooled to its freezing point, 14 K, and partially frozen. The slushy mixture of solid and liquid hydrogen can be pumped just as though it were a homogeneous liquid, and the density of the slush is considerably higher than that of the liquid at its boiling point. R. F. Dwyer and his colleagues at the Linde division of Union Carbide are responsible for much of this work, which is still in the development stage.

The 30,000-pound Centaur, and the 200,000-pound J-2 are the largest hydrogen-oxygen motors which have been flown, but motors as large as 1,500,000 pounds (Aerojet's M-1) are on the way.* All these use electrical ignition. Hydrogen and oxygen are not hypergolic but they are very easily ignited. Gaseous oxygen and hydrogen are admitted to a small pilot chamber, where they are touched off by an electrical spark, whereupon the pilot flame lights off the main chamber. Some work has been done on making oxygen hypergolic with hydrogen, and L. A. Dickinson, A. B. Amster, and others of Stanford Research Institute reported, late in 1963, that a minute quantity (less than a tenth of 1 percent) of O_3F_2 in liquid oxygen would do the job, and that the mixture was stable for at least a week at 90 K (the boiling point of oxygen). O_3F_2, sometimes called ozone fluoride, is a dark red, unstable, and highly reactive liquid produced by an electrical glow discharge in mixtures of oxygen and fluorine at temperatures around 77 K. It has recently been proved that it is really a mixture of O_2F_2 and O_4F_2. However, it doesn't seem likely that electrical ignition of hydrogen-oxygen motors will be supplanted for some time.

The ultimate in hydrogen motors is the nuclear rocket. As we have seen (in the chapter on performance) the way to get a really high performance is to heat hydrogen to 2000 K or so, and then expand it through a nozzle. And that is just what a nuclear rocket motor does. A graphite-moderated enriched

---

* It's a shame that Tsiolkovsky didn't live to see the M-1. It stands twenty-seven feet high, the diameter of the throat is thirty-two inches, and that of the nozzle exit is almost eighteen feet. At full thrust it gulps down almost 600 pounds of liquid hydrogen and a ton and a half of liquid oxygen per second. Konstantin Eduardovitch would have been impressed.

uranium reactor is the energy source, and the hydrogen is the working fluid. (During development, one peculiar difficulty showed up. Hydrogen at 2000 K or so dissolves graphite—it goes to methane—like hot water working on a sugar cube. The answer—coat the hydrogen flow passages with niobium carbide.)

The Phoebus-1 motors, tested at Jackass Flats (lovely name!), Nevada in 1966, with an 1100 megawatt (thermal) reactor, operated successfully at the 55,000-pounds thrust level, with a specific impulse of 760. (Impulses above 850 are expected soon.) The power (rate of change of thermal energy to mechanical energy) was thus some 912 megawatts, which implies that the reactor was working somewhat above its nominal rating. The chamber temperature was about 2300 K.

The Phoebus-2 series nuclear engines, under development, are expected to operate at the 250,000-pounds thrust level; greater than the thrust of the J-2 and the reactor power (thermal) will be about 5000 megawatts. This is twice the power generated by the Hoover dam—and the reactor generating it is about the size of an office desk, An impressive little gadget.

Liquid fluorine work started about the same time as the liquid hydrogen work did. JPL, starting in 1947, was the pioneer. It wasn't particularly available at that time, so they made and liquefied the fluorine on the site, a feat which inspires the respect of anyone who has ever tried to make a fluorine cell work for any length of time. They burned it first with gaseous hydrogen, but by 1948 they had succeeded in firing liquid hydrogen, and were using the latter as a regenerative coolant. And by the spring of 1950 they had done the same with hydrazine. Considering the then state of the technology, their achievement was somewhat miraculous.

Bill Doyle, at North American, had also fired a small fluorine motor in 1947, but in spite of these successes, the work wasn't immediately followed up. The performance was good, but the density of liquid fluorine (believed to be 1.108 at the boiling point) was well below that of oxygen, and the military (JPL was working for the Army at that time) didn't want any part of it.

This situation was soon to change. Some of the people at Aerojet simply didn't believe Dewar's 54-year-old figure on the density of liquid fluorine, and Scott Kilner of that organization set out to measure it himself. (The Office of Naval Research put up the money.) The experimental difficulties were formidable, but he kept at it, and in July, 1951, established that the density of liquid fluorine at the boiling point was not 1.108, but rather a little more than 1.54. There was something of a sensation in the propellant community, and several agencies set out to confirm his results. Kilner was right, and the position of fluorine had to be re-examined. (ONR, a paragon among sponsors, and the most sophisticated—by a margin of several parsecs—funding agency in the business, let Kilner publish his results in the open literature in 1952, but a lot of texts and references still list the old figure. And many engineers, unfortunately, tend to believe anything that is in print.)

Several agencies immediately investigated the performance of fluorine with hydrazine and with ammonia and with mixtures of the two, and with gratifying results. Not only did they get a good performance, but there were no ignition problems, liquid fluorine being hypergolic with almost anything that they tried as a fuel.

Unfortunately, it was also hypergolic with just about everything else. Fluorine is not only extremely toxic; it is a super-oxidizer, and reacts, under the proper conditions with almost everything but nitrogen, the lighter of the noble gases, and things that have already been fluorinated to the limit. And the reaction is usually violent.

It can be contained in several of the structural metals—steel, copper, aluminum, etc.—because it forms, immediately, a thin, inert coating of metal fluoride which prevents further attack. But if that inert layer is scrubbed off, or melted, the results can be spectacular. For instance, if the gas is allowed to flow rapidly out of an orifice or a valve, or if it touches a spot of grease or something like that, the metal is just as likely as not to ignite—and a fluorine–aluminum fire is something to see. From a distance.

But, as is usually the case, the stuff can be handled if you go about it sensibly, and if you want to fire it in a rocket, Allied Chemical Co. will be glad to ship you a trailer truck full of liquid fluorine. That trailer is a rather remarkable device in itself. The inner fluorine tank is surrounded by a jacket of liquid nitrogen, to prevent the evaporation and escape of *any* fluorine into the atmosphere. All sorts of precautions—pilot trucks, police escorts, and what not—are employed when one of those trucks travels on a public road, but sometimes I've wondered what it would be like if a fluorine tank truck collided with one carrying, say, liquid propane or butane.

The development of large fluorine motors was a slow process, and sometimes a spectacular one. I saw one movie of a run made by Bell Aerosystems, during which a fluorine seal failed and the metal ignited. It looked as though the motor had two nozzles at right angles, with as much flame coming from the leak as from the nozzle. The motor was destroyed and the whole test cell burned out before the operators could shut down.

But good-sized fluorine motors have been developed and fired successfully, although none have yet flown in a space mission. Rocketdyne built Nomad, a 12,000-pound motor, burning fluorine and hydrazine, for upper stage work, and Bell developed the 35,000-pound Chariot for the third stage of Titan III. This burned fluorine and a mixture of monomethyl hydrazine, water, and hydrazine, balanced to burn to CO and HF, and to have a freezing point considerably below that of hydrazine. And GE has developed the 75,000-pound X-430 fluorine-hydrogen motor.

Ordin at LFPL, from 1953 on, and then the people at Rocketdyne, in the late 50's and early 60's, investigated the possibility of upgrading the performance of an RP-Lox motor by adding fluorine to the oxidizer (fluorine and oxygen are completely miscible, and their boiling points are only a few degrees apart), and found that 30 percent of fluorine in the lox raised the performance by more than 5 percent, and could still be tolerated (Rocketdyne burned it in an Atlas motor) by tanks, pumps, etc. which had been designed for liquid oxygen. And they got hypergolic ignition, as a bonus. The mixture of liquid fluorine and liquid oxygen is called "Flox," with the usually appended number signifying the percentage of fluorine. For maximum performance the combination should burn (with a hydrocarbon) to HF and CO, which means that Flox 70 is the best oxidizer for RP-1—at least as far as performance goes. The specific impulse of RP-1 and liquid oxygen (calculated at 1000 psi chamber pressure, 14.7 exhaust, shifting equilibrium, optimum O/F) is 300 seconds, with Flox 30 it is 316, with Flox 70 (which balances to CO and HF) it is 343 seconds, and with pure fluorine it drops to 318.

Fluorine is not likely ever to be used for the big boosters—all that HF in the exhaust would be rough on the launching pad and equipment, not to mention the surrounding population—and it's more expensive than oxygen by orders of magnitude, but for deep space work its hard to think of a better combination than hydrogen and fluorine. It's on its way.

The future of ozone doesn't look so promising. Or, to be precise, ozone has been promising for years and years but hasn't been delivering.

Ozone, O_3, is an allotropic form of oxygen. It's a colorless gas, or if it's cold enough, a beautiful deep blue liquid or solid. It's manufactured commercially (it's useful in water purification and the like) by the Welsbach process which involves an electrical glow discharge in a stream of oxygen. What makes it attractive as a propellant is that (1) its liquid density is considerably higher than that of liquid oxygen, and (2) when a mole of it decomposes to oxygen during combustion it gives off 34 kilocalories of energy, which will boost your performance correspondingly. Sänger was interested in it in the 30's, and the interest has endured to the present. In the face of considerable disillusionment.

For it has its drawbacks. The least of these is that it's at least as toxic as fluorine. (People who speak of the invigorating odor of ozone have never met a real concentration of it!) Much more important is the fact that it's unstable—murderously so. At the slightest provocation and sometimes for no apparent reason, it may revert explosively to oxygen. And this reversion is catalyzed by water, chlorine, metal oxides, alkalis—and by, apparently, certain substances which have not been identified. Compared to ozone, hydrogen peroxide has the sensitivity of a heavyweight wrestler.

Since pure ozone was so lethal, work was concentrated on solutions of ozone in oxygen, which could be expected to be less dangerous. The organizations most involved were the Forrestal Laboratories of Princeton University, the Armour Research Institute, and the Air Reduction Co. Work started in the early 50's, and has continued, on and off, ever since.

The usual procedure was to run gaseous oxygen through a Welsbach ozonator, condense the ozone in the emergent stream into liquid oxygen until you got the concentration you wanted, and then use this mixture as the oxidizer in your motor run. During 1954–57, the Forrestal fired concentrations of ozone as high as 25 percent, using ethanol as the fuel. And they had troubles.

The boiling point of oxygen is 90 K. (In working with cryogenics, it's much simpler to think and talk in absolute of Kelvin degrees than in Celsius.) That of ozone is 161 K. On shutdown, the inside of the oxidizer lines would be wet with the ozone-oxygen mixture, which would immediately start to evaporate. The oxygen, with the lower boiling point, would naturally come off first, and the solution would become more concentrated in ozone. And when that concentration approaches 30 percent, at any temperature below 93 K, a strange thing happens. The mixture separates into two liquid phases, one containing 30 percent ozone, and the other containing 75 percent. And as more oxygen boils off, the 30-percent phase decreases, and the 75-percent phase increases, until you have only one solution again—all 75 percent ozone. And *this* mixture is *really* sensitive!

So, after a series of post-shutdown explosions which were a bit hard on the plumbing and worse on the nerves of the engineers, some rather rigorous purging procedures were adopted. Immediately after shutdown, the oxidizer lines were flushed with liquid oxygen, or with gaseous oxygen or nitrogen, to get rid of the residual ozone before it could cause trouble.

That was some sort of a solution to the problem but not a very satisfactory one. Twenty-five percent ozone in oxygen is not so superior to oxygen as to make its attractions overwhelmingly more important than the difficulty of handling it. A somewhat superior solution would be to eliminate the phase separation somehow, and in 1954–55 G. M. Platz of the Armour Research Institute (now IITRI, or the Illinois Institute of Technology Research Institute) had some success in attempting to do this. He showed that the addition of about 2.8 percent of Freon 13, $CClF_3$, to the mixture would prevent phase separation at 90 K, although not at 85 K. Which meant that if you had, say, a 35-percent mixture at the boiling point of oxygen, it would remain homogeneous, but if you cooled it to the boiling point of nitrogen, 77 K, the high concentration, lethal, phase would separate out. W. K. Boyd, W. E. Berry and E. L. White, of Battelle, and W. G. Marancic and A. G. Taylor of Air Reduction, came up with a better answer in 1964–65, when they showed that 5 percent of OF_2 or 9 percent of F_2 added to the mixture completely eliminated the phase separation problem. And their addition didn't degrade the performance, as the Freon would have. Nobody has yet come up with an even faintly plausible explanation for the solubilizing effect of the additives!

One other ozone mixture has been considered—that of ozone and fluorine, which was thoroughly investigated during 1961 by A. J. Gaynor of Armour. (Thirty percent of ozone would be optimum for RP-1.) But the improvement over Flox 70 wouldn't be too impressive, and the thought of what might happen if the ozone in the oxidizer let go on the launching pad and spread the fluorine all over the landscape was somewhat unnerving, and I have heard of no motor runs with the mixture.

For ozone still explodes. Some investigators believe that the explosions are initiated by traces of organic peroxides in the stuff, which come from traces, say, of oil in the oxygen it was made of. Other workers are convinced that it's just the nature of ozone to explode, and still others are sure that original sin has something to do with it. So although ozone research has been continuing in a desultory fashion, there are very few true believers left, who are still convinced that ozone will somehow, someday, come into its own. I'm not one of them.

9

What Ivan
Was Doing

When the Russians moved into Germany, they put the chemists at the Luena works of I.G. Farben to work at propellant research. True, these weren't propellant men, but to the Russians apparently a chemist was a chemist was a chemist and that was all there was to it. ARPA did something similar in this country a good many years later! At first the Germans didn't do much except determine the properties of the known rocket fuels, but when they were sent to Russia in October 1946 (some went to the State Institute of Applied Chemistry at Leningrad, the others to the Karpov Institute at Moscow) they were put to work synthesizing new ones, some to be used neat, some for additives to gasoline or kerosene. For the Soviets, like the Germans before them, were hunting for hypergols, and additives that would make gasoline hypergolic with nitric acid.

And, the nature of chemists and of chemistry being what it is, the paths they took were the same ones we took. They investigated the vinyl ethers, as the Germans had done before them, and then, in 1948, four years before NYU did the same thing, they synthesized and tried every acetylenic that they could think of. In 1948 they tried the allyl amines; Mike Pino at California Research was doing the same thing at the same time. They investigated the tetraalkyl ethylene diamines in 1949, two years before Phillips Petroleum got around to it. And, in 1948 and 1949 they worked over the mercaptans and the organic sulfides, just as Pino was doing. They investigated every amine they could get their hands on or synthesize, and they tried such mixed functional compounds as vinyloxyethylamine. And everything they made they mixed with gasoline—usually a pyrolytic, or high-aromatic type, in the hope that they could get a good hypergolic mixture. They even tried elemental sulfur, in some of their mixtures. But for a long time the most satisfactory fuel for their tactical missiles was the German-developed Tonka 250, mixed xylidines and triethylamine. The second stage of the SA-2 or Guideline (U.S. designations—we don't know theirs) surface-to-air missile used by North Vietnam uses that fuel, along with RFNA.

Home-made hydrazine hydrate (rather than captured German stuff) was available in the Soviet Union by 1948, but there was apparently little interest in hydrazine or its derivatives until about 1955 or 1956, when the Soviet chemists (all the Germans had been sent home by 1950) learned of our success with UDMH. The lack of interest may have been caused by the incompatibility of copper and hydrazine; and their engineers liked to make their motors out of copper, because of its beautiful heat-transfer properties. And, of course, the Russian climate has a tendency to discourage the use of hydrazine. UDMH, now, is one of their standard propellants.

Some work was done with high-strength peroxide, first with captured German material, and, after 1950, with Russian product, but there never was much interest in it, and finally the Navy took over all peroxide work. (It's very useful in torpedoes.)

The nitric acids used in the late 40's and early 50's were a 98 percent WFNA, WFNA containing 4 percent of ferric chloride as an ignition catalyst, and a mixed acid containing 10 percent sulfuric acid. And they had all the troubles with it that we had. They tried organic sulfonic acids—methane sulfonic, methane di and trisulfonic, ethane disulfonic, and ordinary disulfonic acid—as corrosion inhibitors in 1950 and 1951 (two years before California Research tried them) but used them in little more than trace quantities, a percent or so. They didn't work, naturally.

But in spite of the nitric acid troubles, one of the Germans be thought himself of Noggerath's equation relating propellant density to range, and decided to make a few points with his new bosses.[*] He decided that a V-2

---

[*] As a first approximation, the range of a missile is proportional to its boost velocity, squared. And Noggerath related the boost velocity to exhaust velocity and propellant density by the equation:

$$c_b = c \ln (1 + d\varphi),$$

where c_b is the boost velocity, c the exhaust velocity, d the bulk density of the propellants, and φ a loading factor—the total tank volume of the missile, in liters, say, divided by the *dry* mass (all propellants burned) of the missile, in kilograms. So the range depends very strongly upon the exhaust velocity, but upon the density by a logarithmic function which, varies with the loading factor. If φ is very small, as it would be in a plane with JATO attached, the density is almost as important as the exhaust velocity. If it is very large, as an ICBM, the density of the propellants is much less important.

loaded with nitric acid and a really high-density fuel would have a range that would make him a Hero of the Soviet Union, at least, and set out to make that really high density fuel. So he mixed up 10 percent of toluene, and 50 percent of dimethylaniline, and 40 percent of *dibromoethane*. He got a high density all right—something like 1.4—but what those bromines did to the specific impulse was a crime. His Russian bosses, who were not fools, took one horrified look at what he was doing, and immediately took all his chemicals away from him. And four weeks later he was hauled up before a People's Tribunal, tried, convicted, and fined 4000 rubles for, in the words of the court, "Misleading Soviet Science." He was lucky. If I had been on the tribunal he'd have gone to Siberia for ninety years, and the charge would have been Exuberant Stupidity. The Russians were happy when he went back home. With an ally like that who needs enemies?

Other attempts at high-density fuels were made; 8 percent of colloidal aluminum suspended with aluminum stearate in kerosene was one of them. But it froze at −6°, and the investigators lost interest. And they tried various nitro-organics such as nitro-propene—the name alone is enough to scare me to death—as monopropellants, with no success to mention, and, as the Germans had before them, tried to use tetranitromethane as an oxidizer. And blew up a laboratory trying it.

Recently they have been showing considerable interest in mixtures of hydrazine nitrate and methyl hydrazine (like my Hydrazoid N) but whether they intend it for a fuel or for a monopropellant we don't know. Their first ballistic missile the SS-1A (NATO designation), was a carbon copy of the A-4, and burned 70 percent alcohol and liquid oxygen. Liquid oxygen was available in quantity, since the Soviets use the highly efficient and very fast air liquefier designed by Peter Kapitza. The larger missiles, SS-2, "Sibling" of 1954, and the SS-3 "Shyster" of 1956 used the same combination, except that the concentration of the alcohol was 92.5 rather than 70 percent.

But, as you may remember, the U.S. specifications for nitric acid, including the HF inhibitor, were published in 1954. So the next Soviet ballistic missile was a redesigned SS-1A, the SS-1B, or "Scud," and burned kerosene and IRFNA. They presumably used a starting slug—perhaps triethylamine—and the kerosene they use is a high naphthenic type, very similar to RP-1. They prefer this to other types since it is much less liable to coking than, say, a high-olefinic mixture when it is used for regenerative cooling. Suitable crudes are abundant in the Soviet Union. There are two "rocket" grades of IRFNA commonly used in the U.S.S.R., AK-20, containing 20 percent of N_2O_4, and AK-27 containing 27 percent.

From the advent of "Scud," the presence of two design groups in the Soviet Union has been apparent, and the Soviet high command, presumably to keep peace in the family, splits development projects between the two. This procedure is not exactly unheard of in this country, where a contract awarded to Lockheed may be followed by one to General Dynamics.

One group remains wedded to liquid oxygen, and designed the SS-6, SS-8 and SS-10. SS-6, the monstrous 20-barreled beast that lifted Yuri Gagarin and Vostok I into orbit, burned oxygen and the equivalent of RP-1. SS-8, "Sasin," and SS-10 burn oxygen and, apparently a hydrazine-UDMH mixture equivalent to our 50-50.

The other group swears by storable oxidizers, IRFNA or N_2O_4, using the latter in the big strategic missiles which live in steam-heated silos, and the former usually in shorter range tactical missiles which have to cope with the Russian winter. The SS-4 "Sandal" uses IRFNA and apparently a mixture of RP and UDMH (compare U.S. Nike Ajax), while the SS-5 IRBM "Skean" and the SS-7 ICBM burn acid and UDMH. The recently deployed SS-9 ICBM "Scarp," a kissing cousin to the U.S. Titan II, but somewhat larger, burns N_2O_4 with, probably 50–50. There has been some conjecture that it may burn MMH, but that appears unlikely. Fifty-fifty is much cheaper, gives the same performance or a little better, and with a strategic missile you don't have to worry about the freezing point of the fuel. The smaller SS-11 uses the same propellants, and the SS-12, a tactical missile more or less equivalent to the U.S. "Lance," burns IRFNA and RP. (To bring things up to date, the SS-13 is a three-stage solid propellant equivalent to "Minuteman," and the SS-14 is essentially, the two upper stages of SS-13.) The Soviet naval missiles comparable to "Polaris," use IRFNA or N_2O_4 with UDMH or 50–50, or are solid propelled. And the Chinese ballistic missiles under development are based on the SS-3, modified to burn IRFNA and kerosene.

As for more advanced, or "exotic" propellants, the Soviet practice has apparently been more conservative than that of the United States. The Russians did some work with boranes in 1949–1950, but had sense enough to quit before they wasted a lot of time and money. There were some firings with 10 percent ozone in oxygen in East Germany in 1952, but there is no evidence that this work was followed up. Nor is there any evidence of extensive work with halogenated oxidizers. In a long review article on perchloryl fluoride in a Soviet chemical journal recently, all the references were to western sources.* There has been some mention of OF_2, and of the alleged virtues of metal slurries, but nothing to indicate that it amounts to more than words. Nor is there any indication that they have done much with liquid fluorine or with liquid hydrogen,

---

* Of course this *may* mean that they are about to start working with it. Such review articles, in the U.S.S.R., frequently signal the start of a research program.

although it would be surprising, to say the least, if the use of the latter in their space program had not been considered.

In short, the Russians tend to be squares in their choice of propellants. Oxygen, N_2O_4, IRFNA, RP, UDMH and its mixes—that's about the lot. When he wants more thrust, Ivan doesn't look for a fancy propellant with a higher specific impulse. He just builds himself a bigger rocket. Maybe he's got something there.

10

"Exotics"

Fifteen years ago people used to ask me "What *is* an exotic fuel anyway?" and I would answer "It's expensive, it's got boron in it, and it probably doesn't work." I had intended, originally, to entitle this chapter "The Billion Buck Boron Booboo," but decided against it on two grounds. The first was that such a title might conceivably be considered tactless by some of the people who authorized the programs concerned. The second reason is that it would not be completely accurate. Actually, the boron programs did *not* cost a billion dollars. It just seemed that way at the time.

The boranes are compounds of boron and hydrogen, the best known (although there are many others) being diborane, B_2H_6; pentaborane, B_5H_9; and decaborane, $B_{10}H_{14}$. At room temperature the first is a gas, the second a liquid, and the third a solid. Alfred Stock discovered most of the better known boranes between 1912 and 1933, while H. I. Schlesinger, starting about 1930, contributed vastly to the field of borane chemistry, and in particular to the development of synthetic routes.

Boranes are unpleasant beasts. Diborane and pentaborane ignite spontaneously in the atmosphere, and the fires are remarkably difficult to extinguish. They react with water to form, eventually, hydrogen and boric acid, and the reaction is sometimes violent. Also, they not only are possessed of a peculiarly repulsive odor; they are extremely poisonous by just about any route. This collection of properties does not simplify the problem of handling them. They are also very expensive since their synthesis is neither easy nor simple.

But they possess one property which attracted rocket people to them as hippies to a happening. They have an extremely high heat of combustion—gram for gram about 50 percent more than jet fuel. And from 1937 on, when Parsons at JPL had first considered decaborane, propellant men had been considering them wistfully, and lusting after the performance which might, with luck, be wrung out of that heat of combustion.

Nothing could be done about it, of course, until World War II was over. But in 1946 the U.S. Army Ordnance Corps awarded a contract to GE (Project Hermes), to investigate the boranes in depth, and to develop methods of large scale synthesis. The primary objective was not the development of rocket propellants, but the exploitation of the boranes as fuels for air-breathing engines, primarily jets. But the rocket people, as was inevitable with their preoccupations, got involved anyway.

It was Paul Winternitz, at Reaction Motors, who in 1947 made what were probably the first performance calculations on the boranes. He calculated the performance of diborane, pentaborane, and aluminum borohydride, $Al(BH_4)_3$, all with liquid oxygen. Considering the scantiness and general unreliability of the thermodynamic data, not only on these would-be propellants but on their combustion products as well, not to mention the complexity of the calculations (no computers around then, remember!), my admiration of his industry is only equaled by my astonishment at his courage.

At any rate, the numbers that came out at the other end of the calculation, whatever their validity or lack of it, looked encouraging. The next step was to confirm them with motor firings. Diborane (the most available of the boranes) was to be the fuel, and liquid oxygen the oxidizer.

Diborane was the most available of the boranes, but it wasn't exactly abundant. In fact, there were precisely forty pounds of it in existence when RMI started work. So the firings were necessarily at a very low thrust level (perhaps fifty pounds) and were extremely short. At that, as the engineer in charge confessed to me many years later, "Every time I pushed the button I could feel the price of a Cadillac going down the tailpipe!"

The results, not to put too fine a point on it, did not encourage euphoria. The performance was dismally bad—far below theoretical—and solid glassy deposits appeared in the throat (changing its size and shape) and in the diverging (downstream) section of the nozzle. These consisted, apparently, mostly of B_2O_3, but appeared to contain some elemental boron as well. This was a sure indication of poor combustion, and was not encouraging.

Ordian and Rowe, at NASA-Lewis, fired the same combination in 1948, and got much the same sort of results. Nor were the results any better when they used hydrogen peroxide as the oxidizer. The glassy deposits seemed to be as characteristic of borane firings as was the bright green exhaust flame.

The next fuel that RMI tried was the dimethylamine adduct of diborane— not exactly a borane, but a close relative. But when they fired it with oxygen, in 1951, the results were borane results—and discouraging. So were their results with pentaborane, which Jack Gould fired the next year in a fifty-pound thrust motor, using oxygen and hydrogen peroxide as oxidizers. It would be some twelve years before anybody could get good results with that last combination. One with better combustion efficiency was fired by Ordin in 1955—diborane and fluorine. Here, at least, there weren't any deposits in the nozzle—BF_3 is a gas—but the combination was a fiendishly hot one, and very difficult to handle.

The early borane firings weren't, on the whole, too successful, but enthusiasm, hopes, and expectations were all high, and two meetings on boron fuels and would-be fuels were held in 1951 alone. Some awfully dubious chemistry was presented at these meetings—the big breakthroughs in boron chemistry were yet to come—but everybody had a good time and came home inspired to renewed efforts.

And very soon they had the money to make these efforts. Project "Zip" was started in 1952, by BuAer of the Navy. It was designed to carry on from where the Hermes project had left off, and to develop a high-energy, boron-based fuel for jet engines. This was before the day of the ICBM's, the long-range bomber-carrying nuclear bombs was the chosen weapon of deterrence in the cold war, and anything that would increase the range or the speed of that bomber was very much to be desired. The major prime contractors, each with multi-million-dollar contracts, were the Olin Mathieson Chemical Corporation and the Callery Chemical Co., but by the end of the decade many more organizations, propulsion, chemical, academic—you name it—had become involved, either as minor prime contractors or as subcontractors to the primes. By 1956 the program had become so unwieldy that it had to be split, with the Air Force monitoring Olin Mathieson's work and the "HEF" program and Navy's BuAer watching over Callery's "Zip." The trade journals played up the Zip and "Super" fuels (omitting, naturally, the classified chemical details—which, if published, might give some people pause) and legions of trusting and avaricious souls went out and bought boron stocks. And, eventually, lost their shirts.

It soon became evident that in order to attain the desired physical properties (similar to those of jet fuel) the fuels would have to be alkyl derivatives of the boranes. In the end, three of these were developed and put into fairly large-scale production. Mathieson's HEF-2 was propyl pentaborane. Callery's HiCal-3, and Mathieson's HEF-3 were mixtures of mono-, di-, and triethyl decaborane, and HiCal-4 and HEF-4 were mixtures of mono-, di-, tri-, and tetramethyl decaborane. Both −3 and −4 contained traces of unsubstituted decaborane. (The missing numbers represented the fuels in an intermediate stage of synthesis.)

The chemistry of the borohydrides was investigated as it had never been investigated before, process details were worked out on the pilot-plant level, two full-sized production facilities, one Callery, one Mathieson, were built and put on stream, handling and safety manuals were written and published—and the whole thing was done on a crash basis. Never had one poor element been given such concentrated attention by so many chemists and chemical engineers.

And then the whole program was brought to a screeching halt. There were two reasons for this, one strategic, one technical. The first was the arrival of the ICBM on the scene, and the declining role of the long-range bomber. The second lay in the fact that the combustion product of boron is boron trioxide, B_2O_3, and that below about 1800° this is either a solid or a glassy, very viscous liquid. And when you have a turbine spinning at some 4000 rpm, and the clearance between the blades is a few thousandths of an inch, and this sticky, viscous liquid deposits on the blades, the engine is likely to undergo what the British, with precision, call "catastrophic self-disassembly."

All sorts of efforts were made to reduce the viscosity of the oxide, but to no avail. The HEF's and the HiCal's just could not be used in a jet engine. The plants were put on stand-by, and eventually sold for junk. The Zip program was dead, but the memory lingers.

It was by no means a total fiasco. The small fraction of the total cost which went for research added more to the corpus of boron chemistry in ten years than otherwise would have been learned in fifty.[*] One of the most interesting discoveries was that of the "Carboranes," by Murray Cohen, of Reaction Motors, in 1957. The parent compound, $B_{10}C_2H_{12}$, has the structure of a closed, symmetrical, icosahedral cage, and it and its derivatives exhibit a surprisingly high stability against oxidation, hydrolysis, and thermal decomposition. Neff, of Hughes Tool, took advantage of this stability when he attempted to make a monopropellant based on a carborane derivative. (See the monopropellant chapter.) Derivatives may also be useful in high-energy solid propellants, and even, possibly, as high-temperature-resistant plastics.

As far as rocket propulsion itself was concerned, the result of the Zip program was that there were now large stocks of diborane (the starting point for the synthesis of all the boranes and their derivatives) pentaborane, decaborane, and the HEF's and HiCal's available, so that their usefulness as fuels could be investigated on something more than the frustrating fifty-pound level. Aerojet, starting about 1959, worked with HEF-3 and pentaborane, burning them with N_2O_4 or hydrogen peroxide, and Reaction Motors had most of the bugs out

---

[*] Dick Holzmann was at ARPA at the time, and it is due to him that all this chemistry is available, and not buried forever in the files of the contractors and the services. He had all the information collected, heckled Ronald Hughes, Ivan Smith, and Ed Lawless of Midwest Research Institute into putting it together in one volume, and finally edited *Production of the Boranes and Related Research,* which was published by Academic Press in 1967.

of the pentaborane-peroxide system by 1964. With proper injector design the systems could be made to work, and to yield something close to their theoretical specific impulse. And the problem of the solid deposits in the nozzle wasn't too important when the motor was of a respectable size. It didn't arise at all, of course, when a fluorine oxidizer was used. Don Rogillio, at Edwards Air Force Base, in 1962–64 burned pentaborane with NF_3 and with N_2F_4, and got quite a good performance, although, as the combination is a fiendishly hot one, he had a lot of trouble with burned-out injectors and nozzles.

But once pentaborane was made to work, nobody could find any particular use for it. The performance was good, yes, but the density of pentaborane is low—0.618—which militated against its use in a tactical missile. Further, the (oxygen type) oxidizers with which it performed best, peroxide and N_2O_4, had unacceptable freezing points. And if you used nitric acid, you lost a good deal of its performance advantage. And, of course, with any of these oxidizers, the exhaust contained large quantities of solid B_2O_3, and a conspicuous exhaust stream may be undesirable. And if you used a halogen oxidizer, such as ClF_3, the performance wasn't enough better than that of a hydrazine to be worth the trouble. And finally, it was still expensive.

The situation was otherwise with diborane. It couldn't be used in a missile, of course (its boiling point is −92.5°) but might well be used in certain deep-space applications where its low density (0.433 at the boiling point) wouldn't matter. Its natural partner was OF_2 (although ONF_3 would also be suitable) and from 1959 to the present that combination has been under investigation by several agencies, among them Reaction Motors, and NASA-Lewis. The combination is a hot one, and it isn't easy to design injectors and nozzles which will stand it, but the difficulties are far from insurmountable, and an operational system does not seem far away. The combination, by the way, is an unusually hairy one to work with, both propellants being remarkably poisonous, but rocket men usually know how to stay alive, and it hasn't killed anybody—yet.

One thing that might have kept pentaborane in the picture was the advent of BN system, early in 1958. Callery Chemical was the originator of the idea, but within a year *every* propulsion contractor in the country, plus JPL, NASA, and EAFB had got into the act.

This is the idea: Boron nitride, BN, is a white, crystalline solid, with a hexagonal crystal structure like that of graphite.* It is a very stable molecule, with an exothermic heat of formation of some sixty kilocalories per mole. Now, imagine the reaction of a borane with hydrazine.

---

* Carbon, of course, occurs both as graphite and diamond. And some recent work indicates that BN can be had, not only with the graphite structure, but with a diamond-like structure, and as hard, or harder, than diamond itself.

$$B_2H_6 + N_2H_4 \longrightarrow 2BN + 5H_2$$

or

$$2B_5H_9 + 5N_2H_4 \longrightarrow 10BN + 19H_2$$

The heat of formation of the BN would be the energy source, and the hydrogen would comprise the working fluid—dragging the solid BN along with it, of course. Performance calculations indicated that the pentaborane-hydrazine combination should have the astounding performance of 326 seconds, and brought out the even more astounding fact that the chamber temperature should be only about 2000 K–1500 K or so, cooler than anything else with that sort of performance. The thought of a *storable* combination with a performance above 300 seconds, and with such a manageable chamber temperature sent every propulsion man in the country into orbit.

Getting enough pentaborane to work with was no problem, of course, in 1958–59. The Air Force had tons and tons of the stuff, from their Mathieson operation, and hadn't the foggiest idea of what to do with it. So it was practically free for the asking, and everybody leaped into the act, uttering glad cries. Callery, NASA-Lewis, Reaction Motors, and EAFB were some of the first to try the combination—most of them, at first, at approximately the hundred-pound thrust level.

Reaction Motors' experience is typical. Hydrazine/pentaborane was hypergolic, although ignition was a bit hard. Combustion efficiency was ghastly, about 85–88 percent C* efficiency.* And specific impulse efficiency was worse; the engineers considered themselves lucky when they got 75 percent of the 326 seconds the calculations said they should get.

Obviously, the combustion efficiency was the first problem to be tackled, for unless that was brought up to a reasonable figure nothing could be done about the specific impulse—or anything else.

Part of the difficulty stemmed from the fact—soon discovered—that the reaction does not go neatly to BN and hydrogen, as the equations say it should. Instead, some of the boron is exhausted as elemental boron, and the leftover nitrogen combines with some of the hydrogen to form ammonia. This, naturally, does not help performance.

---

* C*, pronounced "see star," is a measure of combustion efficiency. It is derived by multiplying the measured chamber pressure by the area of the throat of the nozzle, and dividing this by the mass flow of the propellants. It comes out in feet per second or meters per second, depending on the system you use. Its theoretical value can be calculated, just as theoretical specific impulse can, and the percentage of theoretical C* that you measure experimentally is a good measure of the completeness of combustion, and of the efficiency of the injector.

Another problem lay in the difficulty of mixing the pentaborane and the hydrazine so that they could react. Hydrazine is a water-soluble substance, and pentaborane is oil-soluble, and the two were remarkably stubborn about getting together. (This led to the BN-monopropellant work, described in the monopropellant chapter.) Additives to the propellants were no help—and everything from hydrazine nitrate to UDMH was tried. To get good mixing you simply have to use a remarkably sophisticated injector. Love, Jackson, and Haberman learned this the hard way, at EAFB, during 1959–60–61. As their thrust level rose from 100 to 5000 pounds, and they laboriously dragged their C* efficiency from 76 percent up to 95 percent, they experimented with no less than thirty different injectors, each one more sophisticated and complicated than the last.

While this was going on, the problems involved in handling pentaborane were still around—and hairy. It was remarkably poisonous, as I have mentioned. And it is hypergolic with the atmosphere, and the fires are brutes to extinguish. If you spray a burning pool of the stuff with water, the fire goes out eventually—if you're lucky. But then the remaining unburned pentaborane is covered with a layer of solid boron oxide or perhaps boric acid, which protects it from the air. And if that crust is broken (which is certain to happen), the fire starts all over again. Even disposing of leftover pentaborane is a problem, but not one I'll go into here. Holzmann's book tells all about it, if you're interested.

Considering all this, I asked some of Rocketdyne's people—Rocketdyne was working closely with EAFB on their BN work—how they managed to live with the stuff. "Oh, it's no problem," they answered. "You just follow the directions in our safety manual!" I asked them to send me a copy of said manual, and in due time it arrived. It would be a misstatement to say that it was the size of the Manhattan phone book, but I've seen a great many municipalities with smaller ones. And even with the help of the manual, one of their rocket mechanics, a little later, managed to get himself hospitalized because of pentaborane.

The final step in the BN work was to scale up to a larger motor, of some 30,000 pounds thrust, and this was done at Edwards during 1961–62–63. (Incidentally, a lot of the work with hazardous propellants has been done at Edwards. It's located in the middle of the Mojave desert, and you don't have to worry about the neighbors. Even if you spill a ton of liquid fluorine—and that's been done there, just to see what would happen—the only thing that's likely to be damaged is the peace of mind of a few jack rabbits and rattlesnakes.) I saw movies of some of the test runs, and they were spectacular, with dense white clouds of solid BN rising two miles into the sky.

The results with the big motor were poor at first—about three-quarters of the theoretical specific impulse—but they improved with injector design, and before the end of 1963 the magic 300 seconds had been reached. (The final injector comprised some *six thousand* carefully drilled orifices! It was not cheap to manufacture.) But the BN system had finally been made to work, and was a success.

The only fly in the ointment was that the system was obsolete at birth. ClF₅ arrived on the scene just as BN succeeded—and the ClF₅–hydrazine combination performs as well as the hydrazine–pentaborane system, is much denser, and much easier to handle, works in a much simpler and cheaper motor, has an invisible exhaust stream—and is cheaper by at least an order of magnitude. Five years of work had been a frustrating exercise in expensive futility. Sometimes rocket men wonder why they ever got into the business.

However, there does seem to be some hope for the BN system, in a rather specialized application. Aerojet, fairly recently (1966–67) has been investigating the usability of the combination in a ram rocket, where the exhausted hydrogen, BN, elemental boron, and ammonia would be burned by the intake air, to give extra thrust, and has found that it works very well indeed in such an arrangement. So perhaps the Edwards people didn't labor entirely in vain.

The borohydrides were related fuels that never quite made it. Here a word of explanation seems to be in order. Borohydrides come in two, or perhaps three types. The first type comprises the alkali metal borohydrides, $LiBH_4$, $NaBH_4$, and so on. These are straightforward ionic salts—white crystalline solids, with no nonsense about them. They are reasonably stable—$NaBH_4$ is *almost* stable in water—and can be handled easily.

Lithium borohydride, as has been mentioned, was tried as a freezing point depressant for hydrazine by Don Armstrong at Aerojet as early as 1948. He found that the mixture was unstable, but nevertheless Stan Tannenbaum, at RMI, tried it again in 1958, with the same results. And then, back at Aerojet, Rosenberg lit on the same mixture in 1965. And *he* found that 3 percent of the borohydride decomposed in 200 days at 69°. All of which gives one a feeling of "this is where I came in."

Sodium borohydride is much more stable than is the lithium salt, and its solution in liquid ammonia is quite stable, and Aerojet fired this, with oxygen, in 1949, but its performance was inferior to that of hydrazine and the work wasn't followed up. And Patrick McNamara, at EAFB, fired a hydrazine solution of the sodium salt with chlorine trifluoride in 1965, but got a performance inferior to that of pure hydrazine.

The second type (the "perhaps") includes ammonium and hydrazinium borohydride, which can be made *in situ* in liquid ammonia or hydrazine, but which would be unstable if isolated at room temperature. Aerojet burned a solution of hydrazinium borohydride in hydrazine (with oxygen) in 1949. I suspect that the mixture was unstable, for nothing more ever came of it.

The third type includes the aluminum and beryllium borohydrides, $Al(BH_4)_3$ and $Be(BH_4)_2$. These are covalent compounds, with unusual bonding, liquids at room temperature, and violently hypergolic with air. Nobody has ever had enough beryllium borohydride all together in one place and at one time for a motor firing, but Armstrong and Young at Aerojet fired aluminum borohydride with oxygen in 1950, and the next year Wilson, also at Aerojet, burned it with liquid fluorine. The results were not sufficiently encouraging to out-weigh the difficulties involved in handling the fuel, and aluminum borohydride lay more or less dormant for some ten years.

Then, starting about 1960, Dr. H. W. Schulz and J. N. Hogsett, at Union Carbide, started the development of the "Hybalines." And—something rare in the propellant business—they did it with company, not government, money. Aluminum borohydride forms a mole for mole addition compound—an adduct—with amines. And these adducts are *not* spontaneously inflammable in the atmosphere, but with reasonable precautions, can be handled without any particular difficulty.

Schulz and Hogsett experimented with dozens of different amines, but the fuel they settled on, as having the best combination of properties, was a mixture of the adducts of monomethyl amine and of dimethylamine. They called it Hybaline A_5. (They also made some adducts of beryllium borohydride. These they called "Hybaline B.") They plugged the Hybalines for some four years, quoting calculated performance figures which were a wonder to behold. The only difficulty was that they assumed—on the basis of certain very doubtful experimental figures—a heat of formation for their mixture of adducts which was incompatible with that generally accepted for aluminum borohydride, and which generated a certain skepticism in their audiences. The question was finally settled when EAFB made a series of full-scale (5000 pound thrust) firings with Hybaline A_5/N_2O_4. And got a maximum of 281 seconds, much less than would have been delivered by, say, ClF_5 and hydrazine. So, by 1964, the Hybalines were finished.

The latest excursion into the realm of the exotic was made by F. C. Gunderloy, at Rocketdyne. He discovered that certain linear polymers of beryllium hydride and dimethyl beryllium, with the chains terminated with BH_3 groups (since the work is classified, I can't be more specific in describing them) were viscous liquids, and worked on them for four or five years. The chemistry involved is interesting for its own sake, but it doesn't appear likely to lead to a useful propellant. The liquids are extremely poisonous, and beryllium oxide, which would be one of the exhaust products if one of them were used as a fuel, is so toxic as to rule out any use in a tactical missile; and there are better fuels for space work. Setting to one side the problems of working with a high-viscosity propellant, beryllium is a comparatively rare and quite expensive material, and there appear to be better uses for it. The development of these compounds would have been an admirable academic exercise well worth several PhD's in inorganic chemistry. As a propellant development program it can be classified only as an unfortunate waste of the taxpayer's money.

So what's ahead for the "exotics?" As I see it, just two things.

1 Diborane will probably be useful in deep space work.
2 The pentaborane/hydrazine, BN system should be very good in ram-rocket and similar systems.

And the people who lost their shirts on boron stocks will have to discover a better way of getting rich from other people's work. For them, my heart does not bleed.

The Hopeful Monoprops

Monopropellants, unlike Gaul, are divided into two parts. Low-energy mono-props are used for auxiliary power on a missile, sometimes for attitude control on a space vehicle (the Mercury capsules, and the X-15 airplane at high altitudes used hydrogen peroxide for attitude control) for tank pressurization and the like. High-energy monoprops, the glamour boys, are intended to compete with bi-propellants for main propulsion.

There haven't been too many of the first sort, and their development has been more or less straightforward. The first, of course, was hydrogen peroxide, used by Von Braun to drive the turbines of the A-4. He used a solution of calcium permanganate to catalyze its decomposition, but later workers at Buffalo Electrochemical Co. (BECCO) found it more convenient to use a silver screen, coated with samarium oxide to do the job. (I'm not sure whether samarium was chosen as a result of a systematic investigation of all of the rare-earth metals, or because the investigator had some samarium nitrate in his stock room.) The leaders in this work were the people at RMI, who were investigating peroxide, at the same time, as the oxidizer for a "super-performance" engine on a fighter plane. They had one interesting monopropellant application of H_2O_2 very much on their minds. This was the ROR, or "Rocket on Rotor" concept, by which a very small—perhaps fifty pounds thrust—peroxide motor was mounted on the tip of each rotor blade of a helicopter. The propellant tank was to be in the hub of the rotor, and centrifugal force would take care of the feed pressure. The idea was to improve the performance of the chopper, particularly when it had to lift off in a hurry. (That means when somebody is shooting at you.) The work on this went on from 1952 to 1957, and was a spectacular success. I've seen an ROR helicopter operating, and when the pilot cut in his rockets the beast shot up into the air like a goosed archangel. The project was dropped, for some reason, which seems a shame. An ROR chopper would have been awfully helpful in Vietnam, where somebody usually *is* shooting at you.

At any rate, peroxide is still used as a low-energy monopropellant, and will probably continue to be used in applications where its high freezing point isn't a disadvantage.

One such application is as a propellant for torpedoes. (After all, the ocean is a pretty good thermostat!) Here it is decomposed to oxygen and superheated steam, the hot gases spin the turbines which operate the propellers, and the torpedo is on its way. But here a little complication sets in. If you're firing at a surface ship, the oxygen in the turbine exhaust will bubble to the surface, leaving a nice visible wake, which not only gives the intended victim a chance to dodge, but also tells him where you are. BECCO came up with an ingenious solution in 1954. They added enough tetrahydrofuran or diethylene glycol (other fuels could have been used) to the peroxide to use up the oxygen, letting the reaction go stoichiometrically to water and carbon dioxide. The water (steam) is naturally no problem, and CO_2, as anybody knows who's ever opened a can of beer, will dissolve in water with the help of a little pressure. That solved the wake problem, but made the stuff fearfully explosive, and brought the combustion temperature up to a level which would take out the turbine blades. So BECCO added enough water to the mixture to bring the chamber temperature down to 1800°F, which the turbine blades could tolerate, and the water dilution reduced the explosion hazard to an acceptable level.

Another low-energy monopropellant was propyl nitrate, first investigated around 1949 or 1950. It was plugged, enthusiastically, in England by Imperial Chemical Industries, who insisted that it was absolutely harmless and nonexplosive. Ha! ERDE (Waltham Abbey) investigated it and its homologues rather extensively, and in this country the Ethyl Corporation and Wyandotte Chemical Co. did the same. The work in England was done on isopropyl nitrate, but in this country, due to a magnificently complicated patent situation, normal propyl nitrate was the isomer used. By 1956, not only Ethyl and Wyandotte, but United Aircraft, JPL, NOTS, Aerojet and the Naval Underwater Ordnance Station (the old Torpedo Station at Newport) were working with it, either as an auxiliary power source or as a torpedo propellant, and either straight or mixed with ethyl nitrate. It was easy to start—either a hot glow bar or a slug of oxygen and a spark plug were enough—burned clean and smoothly, and seemed to be the answer to a lot of problems.

And then it showed its teeth. NPN doesn't go off on the card-gap tester. You can throw it around, kick it, put bullets through it, and nothing happens. But if there is a tiny bubble of gas in it, and that bubble is compressed rapidly—possibly by a water-hammer effect when a valve is closed suddenly—it will detonate—violently. This is known as "sensitivity to adiabatic compression," and in this respect it is at least as touchy as nitroglycerine. It was at Newport that it happened. Somebody closed a valve suddenly, the NPN let go, and the explosion not only did a lot of damage but convinced most rocket people that monopropellant was not for them.

Another low-energy monopropellant that got quite a play starting about 1950 was ethylene oxide, C_2H_4O. It's commercially available, cheaply and in quantity, since it's an important chemical intermediate. It's easy to start—a sparkplug is enough to do it—and decomposes in the reactor to, primarily, methane and carbon monoxide. It has a tendency, however, to deposit coke in the reactor, to an extent which depends upon the nature of the surface of the latter. This effect can be prevented by lining the chamber with silver—the flame temperature is very low—or by adding a sulfur containing compound to the propellant. It is also likely to polymerize in storage, forming gummy polyethylene ethers, which plug up everything. Sunstrand Machine Tool worked with it for several years, using it very successfully to drive a turbine. Experiment Incorporated, Walter Kidde, and Wyandotte Chemical also investigated it, and Forrestal Laboratory, at Princeton, tried it as the fuel of a ram-rocket during 1954 and 1955.

Some work was done on acetylenics, such as methyl acetylene and diisopropenyl acetylene, by Experiment Incorporated, by Air Reduction, and by Wyandotte between 1951 and 1955 but these were never successful as monopropellants—too much coking, even if they didn't decide to detonate.

A monopropellant with better staying power was hydrazine. Louis Dunn, at JPL, investigated it in 1948–51, and it's still with us. It can decompose either to hydrogen and nitrogen, or to ammonia and nitrogen, and the relative importance of the two reactions depends on any number of things: the chamber pressure, catalytic effects, the stay-time of the gases in the chamber, and so on. The reaction is best started by flowing the hydrazine through a catalyst bed into the combustion chamber. Grant, at JPL, in 1953, came up with the first reasonably satisfactory catalyst: iron, cobalt, and nickel oxides deposited on a refractory substrate. The decomposing hydrazine, of course, reduces the oxides to the finely divided metals, which take over the catalytic role after startup. But restarts, if the catalyst bed has cooled, are just about impossible. The Shell Development Company, in recent years (1962 to 1964) has brought out a catalyst which allows restarts—iridium metal deposited on the substrate. But nobody is really happy with it. It's easy to "drown" the catalyst bed by trying to run too much propellant through it, so that you get incomplete decomposition or none at all, and it works very poorly with the substituted hydrazines, which you have to use for low temperature applications. On top of that, iridium is the rarest of the platinum metals and the catalyst is horribly expensive. And just to make it interesting, the major supplier of iridium is the Soviet Union.

Another way to get restarts is to use a "thermal" instead of a catalytic bed. This has a high heat capacity and is insulated against heat loss, so that it will stay hot for some time after shutdown, and will reignite the propellant on restart simply by heating it. For the original start, the bed is impregnated with iodine pentoxide, I_2O_5, or with iodic acid, HIO_3, either of which are hypergolic with hydrazine. But if the period between shutdown and restart is too long——! All that we can say now is that a satisfactory technique for starting hydrazine decomposition is yet to be developed. It's still unfinished business.

During the ten years after World War II, a respectable amount of monopropellant work was going on in England. Not only were the British very much interested in peroxide (both as an oxidizer and as a monopropellant), and in propyl nitrate and its relatives, they were also intrigued with the idea of a monopropellant which could compete with bi-propellants for main propulsion. As early as 1945 they fired the German 80/20 mixture of methyl nitrate and methanol, and came to the regretful conclusion that it was something that just couldn't be lived with, in spite of its respectable performance.

Then the Waltham Abbey people came up with another idea. The "Dithekites" had been developed during the war as liquid explosives, and ERDE thought that they might possibly be good monopropellants. The Dithekites are mixtures comprising one mole of nitrobenzene and five of nitric acid (which makes the mixture stoichiometric to water and CO_2) and a varying percentage of water. D-20 contains 20 percent water. Even with the added water, the mixtures weren't too stable, and the nitrobenzene had a tendency to get nitrated further. But the British tend to be more casual (or braver) about such things than we are in this country, and that didn't deter them appreciably. Nor did another hazard, peculiar to the Dithekites. They are, of course corrosive, and very rough on the human skin, and to make it worse, the highly poisonous nitrobenzene was absorbed rapidly through the damaged tissue into the anatomy of the victim, subjecting him, as it were, to a one-two punch. However, they persevered and fired the things more or less successfully in 1949–50, only to discover that if you put enough water into them to keep them from blowing your head off the performance you got wasn't worth the trouble. End of Dithekites.

Another type of monopropellant that they investigated about this time (1947–48) was based on a mixture of ammonium nitrate and a fuel dissolved in water. A typical mixture was AN-1, consisting of

Ammonium nitrate	26 percent
Methyl ammonium nitrate	50 percent
Ammonium dichromate (combustion catalyst)	3 percent
Water	21 percent

Their performance, unfortunately, was so bad that development was dropped.

In this country, up to 1954, there were two main lines of high energy monopropellant development. One stemmed from the efforts, described in Chapter 3, to reduce the freezing point of hydrazine. As has been related, JPL and NOTS, between 1948 and 1954, had examined mixtures of hydrazine and hydrazine nitrate with a thoroughness which left little to be desired. And it was obvious, of course, that a mixture of hydrazine and hydrazine nitrate would have a better monopropellant performance than straight hydrazine. And when it was tried, which it was by 1950, it was discovered that the obvious was indeed true. There was only one catch. Any mixture which contained enough hydrazine nitrate and little enough water to have a respectable performance was more likely than not to detonate with little or no provocation. So that was not the route to a high energy monopropellant.[*]

Some years later, in the late 50's, Commercial Solvents, working with their own money (which is unusual in the propellant business), and in considerable ignorance of what had been done already (which is *not* unusual in the propellant business) devised a series of monopropellants which were rather similar to the hydrazine mixtures, except that they were based on methyl amine, to which was added ammonium nitrate or hydrazine nitrate or methylammonium nitrate, or lithium nitrate. These were safe enough, but their energy and performance were low.

---

[*] But it seemed to be the way to a liquid gun propellant. Even a low-energy monopropellant has more energy in it per gram than does smokeless ball powder, and a great deal more energy per cubic centimeter. (A liquid is much more closely packed than a heap of small grains.) So, if a liquid propellant were used, either packaged in the cartridge as the solid propellant is, or pumped separately into the gun chamber behind the bullet, it should be possible to get a much higher muzzle velocity without any increase in weight. Hydrazine–hydrazine nitrate–water mixtures have been the usual propellants in the liquid gun programs, although NPN sometimes mixed with ethyl nitrate, has been used at times. These programs have been running, on and off, since about 1950, but have never been carried through. The military demands a weapon, programs are started and run for a few years, then money or interest runs out, and the whole thing ends, only to start all over five or six years later. I've seen three cycles since I got into the business. JPL, Olin Mathieson, Detroit Controls, as well as various Army and Air Force installations, have been involved. The main problems are more in the engineering than in the chemistry.

The other line of high-energy monopropellant work in this country was the development of nitromethane. By 1945, EES, JPL, and Aerojet had worked with it, and had discovered that it could be more or less desensitized by the addition of 8 percent of butanol. JPL did some work with it—finding the optimum injector and chamber design and so on—right after the war, and in 1949 J. D. Thackerey at Aerojet started an intensive study that carried on through 1953. And there was plenty to study!

Ignition was a big problem; it isn't easy to get the stuff going. Aerojet found that you couldn't ignite it with a spark unless a stream of oxygen was introduced at the same time. An ordinary pyrotechnic igniter was useless; it had to be one of the thermite type. One esoteric starting technique they developed was to spray a liquid sodium–potassium alloy into the chamber at startup. That reacted with the nitromethane with sufficient enthusiasm to get things going—but it's not the easiest substance in the world to handle.

Stable and efficient combustion in a reasonably small chamber was another big problem. Aerojet tried dozens of combustion catalyst additives, including such surprising things as uranyl perchlorate, and finally settled on chromium acetylacetonate.

Other additives were tried, to reduce the freezing point and to desensitize the propellant, among them nitroethane and ethylene oxide. They found that the addition of amines, such as aniline, immensely increased the sensitivity, and Fritz Zwicky patented that as an invention in the field of explosives. The final mixture on which they settled comprised 79 percent nitromethane, 19 percent ethylene oxide, and 2 percent chromium acetyl acetonate. They gave it the depressing name of "Neofuel."

Martin and Laurie had been doing similar work for the Canadian Defense Establishment in 1950. Their approach was to try to upgrade the performance of nitromethane or nitroethane or other nitro-alkyls by mixing in a suitable amount of WFNA. (Note the similarity to the Dithekites.) The performance was improved (nitroethane turned out to be the best nitro-alkyl base to start with) but the sensitivity of the mixture made it impossible to live with.

So, in the spring of 1954, the only reasonably high-energy monopropellant that could be used with reasonable safety was Aerojet's "Neofuel." And monopropellant research seemed to be at a dead end.

Then it happened. Tom Rice, at the Naval Research Laboratory, had an idea. He knew that pyridine was extremely resistant to nitration. So, he reasoned that if it were dissolved in WFNA it would probably go to pyridinium nitrate rather than a nitro compound, and then, as the salt, should be quite stable in the acid. And, by varying the amount of pyridine in the acid, he could get any oxidizer-fuel ratio he wanted in the mixture—and should have himself a high energy monopropellant. He tried mixing the pyridine with the acid, got some hissing and sputtering but no violent reaction, and had confirmed his first hypothesis. Then he burned some of his mixture in a liquid strand burner,* and found that it would burn as a monopropellant. He didn't go any further at that time, since he didn't have access to a test stand.

Paul Terlizzi, my then boss, had been visiting NOL, and told me what Tom was doing, just as a piece of gossip. I instantly saw the possibilities, and something else that Paul and, apparently, Tom hadn't thought of. Which was that almost any amine, and not just extremely stable ones like pyridine, could be made into a monopropellant, if its nitrate salt were made first and *then* dissolved in the acid. And God only knows how many amines there are!

I made a few crude performance calculations, and found that trimethylamine should give a performance somewhat better than that of pyridine. Then I had the gang make a small sample of pyridinium nitrate, and another of trimethylammonium nitrate, and mix them up into propellants. This was no trouble—the salts crystallized nicely, and dissolved in the acid with no fuss. We took a preliminary look at the mixtures, and liked what we saw. And then I sat down and wrote a letter to the Rocket Branch in BuAer, asking that I be authorized to look into the whole business. This was early in June, 1954. I should, of course, have waited for official authorization before I did anything more, but as there didn't seem to be any particular reason to observe the legalities, we decided to get going immediately, and to make up a hundred or so pounds of each salt before anybody got around to telling us not to. We had a lot of pyridine around the place, and, for some unknown reason, a tank of liquefied trimethylamine. Plus, of course, unlimited nitric acid, so things went fast.

Making the pyridinium nitrate was easy: Just dissolve the pyridine in water, neutralize with nitric acid, boil off most of the water and crystallize. (But once, during the boiling down process, something went wrong, the mixture started to turn brown and evolve ominous NO_2 fumes, and the whole thing had to be carried hurriedly outdoors and flooded down with a hose!) When we had the dry salt, we dissolved it in the acid in the proportions which would give the best performance, and sent a sample of the mixture down to Tom Rice to try

---

* A liquid strand burner is a gadget which will give you some idea of the burning rate of a monopropellant. It is a pressurized container (bomb), usually with a window. The monopropellant is burned in a narrow (a few millimeters diameter) vertical glass tube. If the tube is not too wide, the propellant will burn straight down like a cigarette, and the rate can be observed and measured. The bomb is pressurized with nitrogen, to pressures similar to those in a rocket combustion chamber, and the burning rate is measured as a function of pressure. It was developed from the strand burner used for solid propellants. Dr. A. G. Whittaker, at NOTS, burned, among other things, a mixture of nitric acid and 2 nitropropane. He was the first to make much use of it. That was in the early 50's.

on his strand burner. It burned better and faster than his had. We investigated the discrepancy, and found that his WFNA had more water in it than ours did.

We named the stuff "Penelope" because we'd been waiting so long for something like it. (Of course, in the original, Penelope did the waiting, but we weren't inclined to be fussy about details.)

The trimethylammonium nitrate went just as easily—except for one small detail. The highly volatile trimethylamine sticks tenaciously to the skin and clothes, and smells like the Fulton Street fish market on a hot Saturday morning (although some of us used a more earthy comparison) and poor Roger Machinist, who had the job of making the salt, was saluted, for some weeks, by people who held their noses with one hand, pointed at him with the other, and shouted, "Unclean, unclean!" We called *that* propellant "Minnie," for reasons which now escape me.

We finally got the authorizing letter from the Rocket Branch at the beginning of September 1954. They insisted that we concentrate our efforts, at first, on the pyridine mixture, so we made up a large batch of Penelope and turned it over to the hardware boys to see what they could do with it.

At this precise juncture Hurricane Hazel dropped the biggest oak tree in New Jersey on me and my MG. We were both out of action for some time and when I got back to work (with a jaw still held together with baling wire) I learned what had happened.

It seems that the engineers had taken a small motor—about fifty pounds thrust—fitted it with a monopropellant injector, and mounted it horizontally on the test stand. They then stuck a pyrotechnic igniter into the nozzle, started it going, and opened the propellant valve. The propellant promptly extinguished the igniter. Two more trials gave the same results. Bert Abramson, who was in charge of the test work, then took an acetylene torch and heated the motor red hot, and opened the prop valve. This time he got ignition, and some half-hearted operation for a few seconds. Inspired to further effort, he crammed about a yard of lithium wire into the chamber, and pushed the button.

Penelope sprayed into the chamber, collected in a puddle in the bottom, and *then* reacted with the wire. The nozzle couldn't cope with all the gas produced, the chamber pressure rose exponentially, and the reaction changed to a high order detonation which demolished the motor, propagated through the fuel line to the propellant tank, detonated the propellant *there* (fortunately there were only a few pounds in the tank) and wrecked just about everything in the test cell. Penelope should have been named Xantippe. She also scared everybody to death—particularly Abramson.

There next occurred what might be called an agonizing reappraisal. It took some months, and then we decided to do what we should have done in the first place. I ordered samples of every reasonably simple amine that I could find on the market, from monomethyl amine to tri hexyl amine. Plus several unsaturated amines, a few aromatics, and some pyridine derivatives.

As soon as the first samples came in I put the gang onto the job of making the amine nitrate salts, which were then made into propellants. At times there would be half a dozen different flasks in the lab each with a different nitrate in it, and all bubbling away at the same time.

That was the case once, when we were all seated around the table in the middle of the lab, having lunch. I glanced up, and noticed that the contents of one flask was turning a little brown. "Who owns that one?" I asked (every man was making a different salt), "Better watch it!" One of them started to get up. The contents of the flask frothed up and then settled, frothed up and settled again, just like a man about to sneeze. I said "Hit the deck!" I got instant obedience when I used that tone, and seven heads met with a crash under the table as the flask and its contents went "Whoosh!" across the top of it. No damage except to the ego of the chemist concerned. But sometimes I wonder how I managed to run that shop for seventeen years without a time-lost accident.

Some of the nitrates couldn't be made into propellants, but would start to react and heat up rapidly when they were mixed with the acid. The unsaturated amines acted this way, as did some of those with long chains, such as the hexyl amines. These were diluted with water and dumped in a hurry. Once we had to call the fire department to do it for us.

The salts varied madly in their physical properties. Some crystallized nicely; others refused to crystallize under any circumstances, and the solution had to be evaporated to dryness over a steam bath, coming out as a fine powder. And some were liquids, even perfectly dry and at room temperature. Monoethylammonium nitrate was one of these—a clear, viscous, slightly greenish liquid. Molten salts are nothing new, but these were the only ones I ever heard of that were liquid at 25°C. I've never found a use for the ethylamine compound, but something with such interesting properties ought to be good for *something!*

But most of them dissolved in the acid without any fuss. I had them made up to $\lambda = 1.00$ (stoichiometric to CO_2 and H_2O), since I expected their sensitivity to be at a maximum at that mixture ratio, and had them card gapped. (We had acquired an old destroyer gun turret—there were dozens of them lying around the place—and set up a card-gap test rig inside it. The idea of the turret was to contain the fragments from the cup holding the test specimen, and to make it possible to find the witness plate after firing.) A goof-off ensign that I was stuck with was supposedly in charge of the card-gap work, but John Szoke, my madly industrious technician, and one of the best men I've ever seen in a lab, did most of the work. And it was a lot of work.

Altogether, he card-gapped about forty different mixtures during that session—and if you can nail down the go-no go point of a single one in less than a dozen shots, you're lucky.

The results of the tests were surprising. First, Penelope and her relatives (derived from pyridine and related compounds) were among the most sensitive of all the mixtures tried—one of them rated some 140 cards. Second, the propellant made from trimethylamine (the one that I'd wanted to try in the first place) was remarkably insensitive—rating about ten cards. And, as the amine samples came in from the manufacturers, a fascinating pattern started to emerge. If you disregarded things like methylcyclohexyl amine, which didn't seem to follow any rules, and considered only those propellants made from straight or branched chain aliphatic amines, the card gap sensitivity appeared to be a powerful function of the structure of the molecule. The longer the chain, the more sensitive was the propellant mixture. A propellant made from propylamine was more sensitive than one made form ethyl amine, and one made from tripropylamine was more sensitive than one made from dipropylamine, which, in turn, was more sensitive than the one made from monopropylamine. And one made from isopropylamine was less sensitive than the normal propylamine mixture.

To say that the explanation of these regularities was not obvious would be to understate the matter. But I proceeded in the way a scientist usually does when he's confronted with a mass of apparently inexplicable numerical data. I worked out an empirical equation relating the card-gap sensitivity to a function, ϕ, which I called "floppiness coefficient," and calculated from the number of carbon chains in the ammonium ion, their lengths, and their degree of branching. (In deriving it, I had to use logarithms to the base *three*, which is something so weird as to be unheard of. Fortunately they canceled out, and didn't appear in the final function!) And from this equation, with the help of the specific heat of the propellant, the size of the ammonium ion, and a few assumptions, I was able to make a guess at the heat of activation or the explosion process. It came out at quite a reasonable figure—some 20 to 30 kilocalories/mole—right in the range of the strength of molecular bonds.

This was interesting, but what was more to the point, my list of candidate propellants was drastically pruned. Starting with thirty-three mixtures, and taking thirty-five cards as an arbitrary sensitivity limit, I had only ten survivors. Some of these I dropped immediately, because the freezing point of the mixture, when made up to the optimum mixture ratio, was too high, or because the dry salt was unstable in storage, or because it was much more expensive than another compound with the same sensitivity.

The final selection was based on thermal stability. Some of the mixtures could be evaporated to the dry crystals over a steam bath, but others, when the acid was almost all gone, would ignite and burn merrily. This was some indication of the relative stability of the propellants, but for more formal—and quantitative—work, we designed and had built a thermal stability tester. This was a small, sealed, stainless steel bomb, with a total volume of approximately

10 cc, equipped with a pressure pickup and a recorder, and with a rupture (or burst) disc failing at approximately 300 psi. The bomb was loaded with 5 cc of propellant and placed in a constant temperature bath, and the pressure buildup was recorded. There was an open "chimney" above the burst disc, which extended above the liquid level of the bath, so that when the disc let go the bath liquid, which was usually old cylinder oil, wouldn't be spread all over the landscape.

In a typical run, a sample was placed in the bath at 100°. In a very few minutes the pressure rose to about 100 psi, and stayed there for about fifteen hours. Then, it started to increase at an accelerating rate, and the burst disc failed at seventeen hours. When we ran a series of runs, at different temperatures, and then plotted the logarithm of the time-to-burst against the reciprocal of the absolute temperature, we got a gratifyingly straight line, from whose slope it was easy to calculate the heat of activation of the decomposition process. (It turned out to be surprisingly close to that derived from the card gap work!)

Anyhow, we found that, other things being more or less equal, secondary amine mixtures were more stable than primary amine mixtures, and that tertiary amine mixtures were the most unstable of all. And of the propellants which had survived our other screenings, that made from diisopropyl amine had the best thermal stability. So it was Isolde. (It was our custom, by this time, to give our monoprops feminine names—like hurricanes. Sometimes the name was vaguely mnemonic of the amine involved—as Beulah for a butyl amine, for instance—and sometimes it had nothing to do with anything. Roger Machinist had been the one to make diisopropylammonium nitrate, and hence had the inalienable right to name it. And he'd been to the opera the night before.)

That was OK with us. Isolde salt was easy to make, crystallized nicely, and was, in general, a joy to work with.

In the meantime, we'd been trying to invent some way of igniting it without blowing up the motor. That wasn't easy. You couldn't set fire to the propellant in the open, even with an oxygen-propane torch. Ordinary pyrotechnic squibs, as we already knew, were useless in a motor. We tried to make some really hot igniters by mixing up powdered aluminum or magnesium, potassium nitrate or perchlorate, and epoxy cement, letting the mess harden in a polyethylene tube, and then cutting off the tube. The results were spectacular. When we lit one of them off (we did it with an electrically heated wire) we got a brilliant white flame, clouds of white smoke, and all sorts of sound effects. We tried them out just outside of the door of the lab, and always had one ready to greet any incautious safety man as he was strolling by. Bert Abramson came in for a demonstration, and when one of them was touched off he tried to extinguish it with a wash bottle. Whereupon the igniter broke in two, with the business end dropping to the floor and chasing him about the lab as everybody cheered. But they wouldn't do the job—a stream of Isolde would put out the fire.

Apparently it just wasn't practical to light the stuff off with an external energy source. It would have to produce its own energy, which meant that we had to develop some source of hypergolic ignition. We didn't want to bother with the plumbing that would be involved if we used a slug of UDMH for instance, to react with the acid in the propellant; that would lead to too much complication. What we wanted was some solid material, which could be placed in the chamber before-hand, and would react with the propellant when it was injected to start the fire going. We tried all sorts of things—powdered magnesium, metallic sodium, and what not. (The candidate was placed in a horizontal one- or two-inch diameter glass tube, the propellant was sprayed in at one end, and the results were monitored with a high-speed camera.) We had no luck for some time, but then we finally hit on a highly improbable mixture that worked—a mixture of lithium hydride and rubber cement. This unlikely sounding mixture was made into a thick dough, spread on a sheet of gauze, and then wrapped around a wooden dowel. The end of the dowel was tapered and screwed into a plug tapped with a $1/8''$ pipe thread. The plug, in turn, was screwed into an appropriately threaded hole in the center of the injector, so that when the propellant was injected it would impinge upon, and react with the ignition mixture. The whole device, some six inches long, was kept in a sealed test tube until it was needed, to protect the LiH from atmospheric moisture. The business end of it was a ghastly corpse-gray, and it was the most obscene looking object I have ever seen—and the rocket mechanics christened it accordingly.*

But it worked. We had our first successful run in January 1956, and by April we had a smooth-running and workable system. We got the best results with a propellant made up with anhydrous nitric acid rather than with ordinary WFNA, and with a salt/acid mixture that gave a λ (the ratio of reducing to oxidizing valences in the propellant) of 1.2. So we called it Isolde 120 A (the 120 referring to the mixture ratio and the A to the anhydrous acid) and wrote our reports. And we had something to report—the highest performing mono-propellant ever fired anywhere by anybody. Combustion was good—we got close to 95 percent of the theoretical performance with a surprisingly small

---

* Since the propellant was named "Isolde" it seemed only reasonable to call the igniter "Tristan." Then somebody pointed out that the missile using the system would naturally be called the "King Mark." But when somebody else added that the advanced model of the missile would of course be the "King Mark II" the engineering officer started muttering wistfully about flogging at the gratings, keel-hauling, and the uses of yardarms, and the "Tristan" idea died an untimely death.

chamber—and we didn't need a fancy (and expensive) injector. In fact, the one we used was made from six commercially available oil-burner spray nozzles costing seventy-five cents each.

Our reports (mine on the development of the propellant and the igniter, methods of analyzing the propellant, card-gap results and so on, and the engineering report on the motor work) came out together in November 1956, but everybody in the business had a pretty good idea of what we had done by June. And then all hell broke loose.

Everybody and his uncle wanted a piece of the action, and wrote a proposal to one of the three services, for a research program on monopropellants. RMI, right next door to us, and intimately acquainted with our work, was first off the mark, in March 1956, when they got a Navy contract to develop "Superior Liquid Monopropellants," but the others weren't far behind. Wyandotte Chemical had a Navy contract by September, Phillips Petroleum and Stauffer Chemical got into the act early in 1957, and by 1958 Pennsalt, Midwest Research, Aerojet, and Hughes Tool had joined them. In addition to these, all of whom were trying to brew up new propellants, several organizations, including GE, were motor-testing propellants that others had developed, and were trying to apply them to tactical systems. It was a busy time.

Reaction Motors (before long they had not only a Navy monopropellant program, but an Army contract as well) tried two approaches. One was to dissolve a fuel in an oxidizer, and the other was to produce a single-compound propellant, the nitrate or the nitramine of an energetic radical. Propargyl nitrate, propargyl nitramine, glycidyl nitrate, 1,4 dinitrato 2 butyne, and 1,6 dinitrato 2,4 hexadyne are typical of the monstrosities they produced. (Reading the names is enough to dampen a propellant man's brow!)

I don't believe that they ever made enough of any of them to do a card-gap, but the results of certain other tests were enough to make one a bit cautious. Joe Pisani phoned me from RMI one day late in 1958, asking me if I'd do a thermal stability run on a sample of propargyl nitrate. I replied that I'd be glad to, but that he'd have to replace anything that got busted, since I didn't trust the stuff. So he sent his sample up to us. It was only 3 cc (we usually used 5) but maybe we were lucky at that. John Szoke heated the oil bath up to 160°C (the temperature that we used then for routine tests) loaded the sample into the bomb, lowered the bomb into the bath, and scurried back into the lab, closing the door behind him. (For obvious reasons, the setup was outdoors and not in the lab.) He turned on the recorder and watched. Nothing happened for a while. The pressure rose slowly as the sample warmed up, and then seemed to stabilize.

And then it let go, with an ear-splitting detonation. Through the safety glass window we saw a huge red flare as the oil flashed into flame, only to quench immediately as it hit the ice-cold concrete. We cut everything off, and went out to survey the damage. The bomb had fragmented; the burst disc just couldn't rupture fast enough. The pressure pickup was wrecked, as was the stirrer. The cylindrical stainless steel pot which had held the oil had been reshaped into something that would have looked well under a bed. And the oil—it had been old vacuum pump oil, black and filthy. It had hit the concrete floor of the test area, the wall of the building, and everything else in reach, and had cleverly converted itself (the temperature was well below freezing) into something resembling road tar. I got on the phone.

"Joe? You know that stuff you sent me to test for thermal stability? Well, first, it hasn't got any. Second, you owe me a new bomb, a new Wianco pickup, a new stirrer, and maybe a few more things I'll think of later. And third (*crescendo* and *fortissimo*) you'll have a couple of flunkies up here within fifteen minutes to clean up this (—bleep—) mess or I'll be down there with a rusty hacksaw blade. . . ." I specified the anatomical use to which the saw blade would be put. End of conversation.

And it was the end of the propargyls and their relatives. Washington told Reaction to forget that foolishness and start working on the N-F compounds instead. That story will be told a little later.

The other approach to a monopropellant at RMI was taken by Stan Tannenbaum, who tried mixtures of inert (he hoped) oxidizers and fuels. This was bathtub chemistry, involving little or no synthesis, but requiring strong nerves. It had the advantage, of course, that the stoichiometry could be adjusted ad lib, and wasn't constrained, as in the one-component monoprops, by the nature of the molecule. And the idea wasn't exactly new. The French, during World War I, had employed aerial bombs filled with a mixture of N_2O_4 and benzene. (The stuff was so touchy that the two liquids weren't mixed together until the bomb had been dropped from the plane!) And, incidentally, some years before I got into the monopropellant business, a hopeful inventor had tried to sell me this same mixture for a monoprop, averring that it was as harmless as mother's milk. I didn't buy it.

Stan worked with N_2O_4 and with perchloryl fluoride. He found that he could mix bicyclooctane or decalin in N_2O_4 without immediate disaster, but that the mixture was too touchy to live with. He tried tetramethyl silane too, in the hope (unrealized) that it would be safer, but finally and regretfully came to the conclusion, late in 1959, that you could not make a practical monopropellant based on N_2O_4. Howard Bost, at Phillips Petroleum, who had been working with mixtures of N_2O_4 and neopentane or 2,2 dinitropropane, came to the same conclusion at about the same time. And if any more evidence were needed, the card-gap values for various N_2O_4-hydrocarbon mixtures, as determined by McGonnigle of Allied Chemical, furnished it. N_2O_4-fuel mixtures are not useful monoprops.

He didn't have any more luck with perchloryl fluoride. He tried first mixing it with amines, but found that if they dissolved at all they immediately reacted with the oxidizer. He *could* dissolve hydrocarbons or ethers, but the mixtures were touchy and too dangerous to handle. (The same discovery was made at GE, when a mixture of perchloryl fluoride and propane detonated, seriously injuring the operator.) So that approach, too, was hopeless. Nor was it a good idea to try to mix N_2F_4 with monomethyl hydrazine, as he discovered early in 1959!

If Tannenbaum's mixtures were bad, that proposed at a monopropellant conference in October 1957 by an optimist from Air Products, Inc., was enough to raise the hair on the head of anybody in the propellant business. He suggested that a mixture of liquid oxygen and liquid methane would be an extra high-energy monopropellant, and had even worked out the phase diagrams of the system.[*] How he avoided suicide (the first rule in handling liquid oxygen is that you never, *never* let it come in contact with a potential fuel) is an interesting question, particularly as JPL later demonstrated that you could make the mixture detonate merely by shining a bright light on it. Nevertheless, ten years later I read an article seriously proposing an oxygen-methane monopropellant! Apparently junior engineers are allergic to the history of their own business.[**]

The work done at Wyandotte by Charlie Tait and Bill Cuddy wasn't quite as hairy as that performed at Air Products, but it approached it closely enough to satisfy a reasonably prudent man. For one thing, Bill, like Joe Pisani, synthesized some really fancy organic nitrates, such as 1,2 dinitratopropane, and nitratoacetonitrile, and as might have been expected, discovered that nobody in his right mind would try to use them as propellants. For another, he examined the possibility (admittedly slight) of using alkyl perchlorates, such as ethyl perchlorate, $C_2H_5ClO_4$, as monoprops. I read in a Wyandotte report that they intended to do this, and phoned Bill to read to him what Sidgwick, in "Chemical Elements and their Compounds," had to say on the subject of the ethyl compound.

---

[*] His idea was to set up a liquid oxygen plant alongside a natural gas well, tank up your ICBM on the spot, and push the button.

[**] Sometime later, Irv Glassman, of the Forrestal Laboratories, conceived of an interesting and entirely different type of cryogenic monopropellant. The idea was to use a mixture of acetylene and excess liquid hydrogen. When they reacted, the product would be methane, which, with the excess hydrogen, would be the working fluid, while the heat of decomposition of the acetylene plus that of formation of the methane would be the energy source. Considering the theoretical performance, the chamber temperature would have been remarkably low. The idea, however, has not yet been tested experimentally.

"Hare and Boyle (1841) say [Sidgwick wrote] that it is incomparably more explosive than any other known substance, which still seems to be very nearly true. . . . Meyer and Spormann (1936) say that the explosions of the perchlorate esters are louder and more destructive than those of any other substance; it was necessary to work with minimum quantities under the protection of thick gloves, iron masks [Ha, there, M. Dumas!], and thick glasses, and to handle the vessels with long holders." But Cuddy (presumably investing in leather gloves and an iron mask first) went ahead anyway. He told me later that the esters were easy enough to synthesize, but that he and his crew had never been able to fire them in a motor, since they invariably detonated before they could be poured into the propellant tank. It is perhaps unnecessary to add that this line of investigation was not further extended.

A system on which they worked for more than two years was based on a solution of a fuel in tetranitromethane—the stuff that had meant nothing but trouble to everybody who had ever had anything to do with it. And Bill and Charlie had their troubles.

One fuel they tried was nitrobenzene. It dissolved nicely in the TNM, to make a propellant with the proper oxygen balance, and the solution seemed reasonably stable. But when they card-gapped it, they found that its sensitivity was over 300 cards. (In my own work, I flatly refused to have anything to do with anything with a card-gap figure much over 30.) Acetonitrile, which they chose as a fuel (they had calculated the performances of dozens of the possibilities, and had tried a few of them) wasn't quite as bad as the nitrobenzene, but it was bad enough. But about this time some of the people in the monoprop business, when accused of producing something which was insanely hazardous, would answer blithely, "Sure, I know it's sensitive, but the engineers can design around it." (The engineers took a dim view of this.)

So they went ahead anyway, and actually managed to fire the stuff in a micro-motor. Most of the time. Sometimes, generally, and embarrassingly, when they were demonstrating it to visitors, it would let go with a frightful bang, demolishing the motor and the instrumentation, and scaring everybody half to death. Tait and Cuddy sweat blood, but they were never able to make the TNM mixtures into reliable propellants, and late in 1958 the thrust of their work shifted to the amine nitrates.

If Tait and Cuddy were fighting a lost cause, Jack, Gould, at Stauffer, must have been smoking Acupulco Gold. His investigations were pure fantasy, to be described properly only by Lewis Carroll. He had a Navy contract to develop "High Energy Monopropellants," and his efforts in that direction challenge belief. The most sensible thing he tried was to dissolve NH_3 in NF_3. Both are quite stable compounds, and he might have come up with a high performing and reasonably safe propellant. Unfortunately, the ammonia wouldn't dissolve in the NF_3 to any extent. Otherwise:

He tried to make nitronium borohydride, NO_2BH_4, and failed. (The idea of a stable salt with an oxidizing cation and a reducing anion isn't very plausible on the face of it.)

He tried to mix pentaborane with nitro-ethyl nitrate. They exploded on contact. (The NEN by itself card-gapped at about 50 cards.)

He tried to mix NF_3 and diborane. They reacted.

He tried to mix NEN with amine derivatives of various boranes. They reacted or exploded.

And so on indefinitely. And in every quarterly report he would list a lot of hypothetical compounds, like di-imide, H—N=N—H, which would be beautiful propellants if you could only make them. Finally the Rocket Branch, fed up, told him to quit that sort of foolishness and to work on the NF systems instead. Which, starting about the end of 1958, is what he did.

While all this was going on, the amine nitrate monopropellants were very much in the picture—but NARTS was not alone there with them. GE jumped in and started motor work, trying to use Isolde in a novel self-pumping motor that they were developing. (They blew up their setup, which goes to show that it isn't a good idea to try to develop a new type of motor with an experimental propellant. One unknown at a time is plenty to worry about!)

Even before the Isolde report was published, Bost and Fox at Phillips Petroleum had made nitrate salts of some of their bi-tertiary amines, and dissolved the salts in nitric acid to come up with AN propellants of their own. They discovered, however, that their thermal stability was extremely poor, which agreed with our own experience with tertiaries. They also discovered that their di-nitrate solutions were extremely viscous, as we had learned, very early in the game, when we tried to make a propellant from ethylene diamine.

At NARTS, the engineers plowed ahead, trying Isolde in high pressure motors—1000 instead of 300 psi chamber pressure—and as a regenerative coolant. It *could* be used that way, but the process was somewhat precarious. You had to shut down with a water flush through the system, or the propellant left in the cooling passages of the still-hot motor would cook off and probably blow up the works.

This was all very commendable, but not very interesting to anybody except a hardware merchant. Which I was not. So I decided to see whether quaternary ammonium nitrates would make better propellants than secondary ammonium nitrates. We had never investigated the quaternaries, since they were comparatively difficult to make, and there was no a priori reason to believe that they would be any better or any worse than Isolde. But there was only one way to find out.

We had a little tetramethyl ammonium hydroxide in the laboratory, so I made it into the nitrate salt—it crystallized beautifully—and had it made up into propellant. We didn't have enough for card-gap work, but we tried it in the thermal stability tester. And got a shock.

It was incredibly stable. When Isolde cooked off in fifty minutes at 130°, the new stuff just sat there. And at 160° it would stand up for more than a week with nothing happening. (Isolde lasted two minutes at 160°.)

This was exciting, and we looked around for a way to make more of the stuff. Tetramethyl ammonium nitrate wasn't commercially available—there was no reason why anybody would have wanted it before—but the chloride was, and I ordered enough to convert into a reasonable amount of the salt we wanted. The conversion was easy enough, even though we used up a lot of expensive silver nitrate doing it (we later went to the trouble of reclaiming the silver, so we could use it over and over) and we soon had enough propellant to do card-gap tests. And our new propellant at $\lambda = 1.2$ had a sensitivity of about five cards, which meant that the shock wave pressure that was needed to set if off was more than twice the pressure that would detonate Isolde.[*] Immensely cheered, we christened her with the obvious name "Tallulah" (as being practically impervious to shock) and continued on our way. This was early in 1957.

The only trouble with Tallulah was that when it was mixed up to $\lambda = 1.20$, the freezing point was much too high—about –22°. (Such a beautifully symmetrical ion can hardly be restrained from crystallizing.) So we next tried the ethyl trimethyl salt ("Portia," but don't ask me to explain the convoluted line of reasoning that led to that name!) and the diethyl dimethyl ammonium nitrate. ("Marguerite," and don't ask me to explain that one either.) Portia didn't quite make it—it would meet the freezing point specifications at $\lambda = 1.10$, but not at

---

[*] About this time I got curious as to whether or not the structure-sensitivity relationships I had found in the amine nitrate monoprops applied to other systems, particularly since McGonnigle had remarked to me that straight chain hydrocarbons in N_2O_4 were more sensitive than branched chains. I knew that Tallulah at $\lambda = 1.0$

$$CH_3$$
with the fuel ion structure $CH_3-\overset{\overset{\displaystyle CH_3}{|}}{\underset{\underset{\displaystyle CH_3}{|}}{N}}{}^+-CH_3$ card-gapped at some eight cards, while

the propellant with the isomeric fuel ion $NH_3^+-CH_2-CH_2-CH_2-CH_3$ gapped 58—a difference of fifty cards due to structure alone. So I got some normal pentane, $CH_3-CH_2-CH_2-CH_2-CH_2-CH_3$ and some neopentane

$$CH_3$$
$CH_3-\overset{\overset{\displaystyle CH_3}{|}}{\underset{\underset{\displaystyle CH_3}{|}}{C}}-CH_3$ and had them both made up with N_2O_4 to $\lambda = 1.0$, and had them

card-gapped. And the normal pentane had a sensitivity of about 100 cards, and the neopentane, 50 cards. Again a difference of 50 cards, which meant that the ratio of the critical shock wave pressures due to structure was the same in both systems. I was fascinated by this coincidence, but never had the chance to carry the work any further. It's recommended to the attention of some future investigator.

1.20, and it crystallized poorly and was rather hygroscopic. Marguerite met the freezing point specifications all right, but it crystallized very badly, and was so hygroscopic as to be practically unusable.

These salts we had made for us by outside manufacturers who had the pressure equipment which we did not have, and which is practically indispensable in making quaternaries in any quantity. John Gall at Pennsalt, and Dr. Phyllis Oja of Dow were both remarkably helpful here in talking their respective pilot plants into making the things and absorbing the financial loss involved. (We got the salts for much less than it cost the manufacturers to make them.)

If Marguerite had poor physical properties, a more compact and symmetrical isomer might do the job, and trimethyl isopropyl ammonium nitrate was the next thing we tried. That was fine—good freezing point, excellent thermal stability, a little more sensitive than Tallulah on the card-gap, but not enough to matter, and physical properties that made it a joy to handle. We called it "Phyllis." (After all, when a lady talks her employers into making 150 pounds of a completely unheard of salt for you, and then doesn't charge you anything for it, on the grounds that the paper work would cost more than the stuff is worth, the least a gentleman can do is to name it after her!)

At the end of 1957, Phyllis looked like the best bet, but we kept on looking. All through 1957, and for three years more, we scurried about, rounding up likely looking amines, quaternizing them and checking them out. We'd usually make enough at first to check the thermal stability and the melting point, and then, if it passed these tests (most of them didn't) make a lot big enough for card-gap work. And if a candidate showed up well there, it was time to look for somebody who would make enough for motor work.

In January, 1958, Bost and Fox, of Phillips, with a new Air Force contract, returned to the monopropellant business with a splash. Phillips, of course, had all the fancy equipment that a man could desire, and they could work fast. For instance, if they wanted the fuel-ion

$$\underset{\underset{\displaystyle C}{|}}{\overset{\overset{\displaystyle C}{|}}{C\!-\!\overset{+}{N}\!-\!C\!-\!C\!-\!\overset{+}{N}\!-\!C}}$$

(hydrogens, as usual omitted for simplicity) they would simply react ethylene chloride with trimethylamine, in almost any solvent, under pressure, and have what they wanted. We envied them their equipment and cursed the affluence of the petroleum industry. Anyhow, they synthesized about a dozen different quaternary amine nitrates, dissolved them in nitric acid, and checked out their properties. They did some work with perchlorates, but found that they were entirely too sensitive; and some with N_2O_4 and N_2O_4-H_2O mixtures, but found that the nitrate salts weren't sufficiently soluble in N_2O_4 to make a propellant, and that when enough water was added to make them soluble they lost too much energy. So they resigned themselves to investigating the same sort of systems that we were working on. And for a year or so our two programs went along more or less in parallel—they working on double ended propellants, we working on single enders.

A newcomer in the field was J. Neff, of Hughes Tool. (Yes, Howard Hughes' company.) Early in 1958, armed with a Navy contract and more than the usual complement of optimism, he started work on the development of a boron-based nitric acid monopropellant. The program lasted about a year and a half, and while it didn't lead to a useful propellant, it involved some interesting chemistry. His most nearly successful approach was based on the carborane structure, in which two carbon atoms take their place with the ten borons in the open basket decaborane structure to form a twelve-atom closed icosahedral cage. (See the boron chapter.) To one or both of these carbons he would attach a dimethylaminomethyl or dimethylaminoethyl group, then make the nitrate salt of the result, and dissolve the salt in nitric acid. In some cases he managed to get away with it, although ignition was likely if he mixed the components too rapidly. But his solutions were unstable; they would evolve gas if they were warmed up a bit, or would separate into two layers, or do something else to emphasize that this was not the way to make a monopropellant. His work never got to the stage of motor testing.

Nor did Aerojet's monopropellant work for the Air Force. Late in 1958, M. K. Barsh, A. F. Graefe, and R. E. Yates started investigating certain boron ions, such as $[BH_2(NH_3)_2]^+$, with the intention of making the nitrates and dissolving them in nitric acid. There was a whole family of these ions, sometimes with hydrazine in place of the ammonia, some containing more than one boron atom, and so on. The chloride of the one shown above can be obtained by milling together in a ball-mill lithium borohydride and ammonium chloride. Max Barsh and Co. called these ions the "Hepcats," from High Energy Producing CATions. (My deplorable habit of giving propellants fancy names was apparently catching!) They made some attempt to synthesize the aluminum analogs of some of the ions, without any notable success. But unfortunately, by the end of July 1959, they had discovered that the Hepcats weren't stable, even in water, let alone nitric acid or N_2O_4. End of the Hepcats.

As I have said, Phyllis seemed to be the most promising AN around at the beginning of 1958, and by the end of that year it, as well as Tallulah, had been fired successfully by NARTS and by Spencer King, of Hughes Tool. Howard Bost's "Ethane" had also been fired, and, as far as performance was concerned, there didn't seem to be much to choose among them. I don't think that anybody ever actually fired either Portia or Marguerite. In most of these monopropellant tests, ignition was by a UDMH slug. This was more complicated, of course, than using our "Tristan" igniter, but it was considerably more reliable for test stand work.

The next development was touched off by John Gall, of Pennsalt, who, in the summer of 1958, sent me samples of two amine nitrates for evaluation. The ions are shown below.

There was some delay—one of my gang goofed and made the test propellants up in the wrong proportions, so we had to ask for another sample—but we finally managed to run off thermal stability tests. The tertiary ammonium nitrate, of course, was no good, but the quaternary appeared to be at least as stable as Tallulah. It then occurred to me that it might be interesting to compare the three different, but very similar, ions shown below, which might be thought of as made of two Tallulah ions joined by one, two, or three links respectively.

Mike Walsh, of our laboratory, announced our intention of doing this at an ARS meeting in the middle of September, and we started in at once. The nitrate of the first ion was easily available: It was Howard Bost's "Ethane" salt. For the third, we got another and larger sample from John Gall. The second had never been made, but Jefferson Chemical Co. made N,N′ dimethylpiperazine, and it wasn't any trick at all to quaternize that and get the salt he wanted. Anyway, we made up the propellants, and tried them in the thermal stability tester at 160°. Number 1 lasted for a bit more than two hours. Number 2 lasted about two minutes. And number 3 just sat there, doing nothing, until we got bored with the whole business and cut off the test after three days.

This was verrrry interesting—apparently we had something even tougher than Tallulah. So we made it up to $\lambda = 1.2$, and card-gapped it. And discovered, to our astonishment, that it did not detonate even at zero cards. This was more than interesting—it was sensational. The freezing point was bad, $-5°C$, but we figured that we could get around that somehow, and it didn't dampen our enthusiasm.

It had to have a good name. Nobody was going to call it by its formal title, 1,4, diaza, 1,4, dimethyl, bicyclo 2,2,2, octane dinitrate, that was for sure. The ion had a nice symmetrical closed cage structure, so I called it "Cavea" (which, after all, has a vaguely feminine sound) after the Latin for cage. Nobody objected—it was easy to remember and to pronounce, although lots of people asked me what it meant!

We went through the usual routine, burning the salt in the calorimeter to get its heat of formation, measuring the heat of solution of the salt in the acid (these two so we could make decent performance calculations) measuring the density and the viscosity of the propellant as a function of temperature, and all the rest. And everything was fine except that freezing point.

One valiant attempt to remedy the situation went on for months, and got exactly nowhere. I reasoned that a single ended ion, such as,

would result in a propellant with a freezing point that would suit anybody. The difficulty lay in getting the nitrates of either of the ions in question. I shopped around for months before I found an outfit that would—or could—make me a sample of the first one. And when it finally came in and was made into a propellant, the thermal stability was abysmally bad. Spurge Mobley, in my own laboratory, synthesized the other (it was the very devil of a job and took him weeks), and it, too, as a propellant, had an impossibly bad thermal stability. Oh, well, it was a good idea, anyway.

In the meantime we (and BuAer) were in a hurry to get Cavea into a motor. So we wanted a large amount of Cavea salt—fast.

I knew that John Gall's people had made it by methylating triethylene

diamine and at the beginning of December he told me that

the latter came from Houdry Process Co. It was used for a polymerization catalyst and sold under the peculiarly repulsive trade name "Dabco." In the meantime, he could furnish me with Cavea salt for about seventy dollars a pound, in ten pound lots. I put in an order, but I wasn't entirely convinced that the salt couldn't be made more cheaply than by reacting methyl iodide (which is quite expensive) with triethylene diamine, and then metathesizing with silver nitrate.

Maybe I could do the synthesis differently. Instead of starting with

and putting the carbons on the ends to get ,

might it not be possible to start with Jefferson Chemical's N,N′ dimethylpiperazine,

and plug in a two carbon bridge to get the same thing? I could use ethylene bromide (which is a whole lot cheaper than methyl iodide) as the bridging agent, and would come up with the bromide rather than the iodide salt.

We tried it, and the reaction worked beautifully, giving us about a 95 percent yield on the first try. The next thing to do was to find a cheap way to convert from the bromide to the nitrate.

I knew that it was quite easy to oxidize the bromide ion to free bromine, but that it was considerably tougher to force it up to the bromate. And I was pretty sure that nitric acid would do the first and not the second. If I added the Cavea bromide to fairly strong, say 70 percent, nitric acid, the reaction should go

$$2Br^- + 2HNO_3 \longrightarrow 2HBr + 2NO_3^-$$

and then

$$2HBr + 2HNO_3 \longrightarrow 2NO_2 + 2H_2O + Br_2.$$

And if I ran a stream of air through the mixture, to blow out the bromine and the NO_2, I should be left with a solution of Cavea nitrate in fairly dilute nitric acid.

We tried it, and it worked. But we discovered that if we added the salt to the acid too fast, or let the bromine concentration build up, we got a brick-red precipitate of Cavea tribromide—the salt of the anion, and it took hours of blowing before *that* would dissociate and release the bromine. I had heard vaguely of the possibility of such an anion, but that was the first time I ever saw one of its salts, Anyway, we dried out the nitrate over a steam bath, recrystallized it from water (it crystallized in beautiful hexagonal crystals) and had our Cavea salt by a simple route that didn't require any expensive reagents. By this time it was the middle of February 1959, and we had learned that Howard Bost, working with his di-quaternaries, had independently hit upon Cavea, and had, like us, decided that it was the best propellant to work with. So now our two programs had converged completely. This was emphasized at a symposium on the AN monopropellants, held at NARTS on the first and second of April. Never have I met such unanimous agreement in such a high-powered group. (The nineteen guests comprised eighteen PhD's and one drunken genius.) And we were all convinced that the future belonged to Cavea, possibly with some structural modification that would get us a better freezing point.

One new development, however, was reported by Dr. Wayne Barrett, of W. R. Grace Co. He had methylated UDMH, to get the ion,

$$CH_3 - \overset{\overset{\displaystyle CH_3}{|}}{\underset{\underset{\displaystyle CA_3}{|}}{N}}{}^+ - NH_2$$

and had made a monoprop with the nitrate of that. Besides the trimethyl compound shown, he had also made the triethyl and the tripropyl, and was so new and innocent in the propellant business that he didn't even start to run when, as he mixed up the propyl salt with the acid, the stuff had started to warm up and give off NO_2 fumes! Anyway, I had my gang immediately make up some of his propellant and give it the treatment. We learned about it on Thursday the second of April, had the synthesis and purification completed and the methyl propellant made up by Tuesday the seventh, and on the eighth saw it wreck our thermal stability tester all over again. (It sat there quietly for fourteen minutes and then detonated—but violently.) I phoned Barrett, and warned him against his brain-child, but he decided to go ahead anyway and repeat our stability test. After some weeks he phoned me and reported that his sample had lasted for seventeen minutes before blowing up the place, and did I consider that a good check!

In the meantime, I had been taking steps toward getting Houdry and Jefferson interested in manufacturing Cavea salt. I had phoned them both on February 19, describing the salt I wanted, and asking if they were interested in bidding on a hundred pounds of it. At Houdry, apparently everybody took off on cloud nine. (This was early in 1959, remember, the cold war was on, everybody was excited over missiles and space, and apparently everybody was convinced—falsely—that there was a lot of money to be made in the rocket propellant business.) Anyhow, they phoned me back several times, and when I got home that evening they had the cleaning lady on the phone. And the next day they had their director of research up from Philadelphia to talk to me. Before I got through with them I told him that Jefferson Chemical was in on the bidding, too, and hinted that perhaps Jefferson could make the stuff cheaper than Houdry could.

The response of Jefferson Chemical was not quite so hysterical, but enthusiastic enough. I got their director of research in Houston, Dr. McClellan, and described the dimethyl piperazine-ethyl bromide reaction to him—he didn't really believe it until he tried it himself—and asked him what he could do. I also hinted to him that perhaps Houdry could give me a better price than he could. This is the procedure known as playing both ends against the middle.

Both companies delivered preliminary samples for approval within a month, and I discovered that McClellan had come up with an interesting method of getting rid of the bromide. He would acidify the Cavea bromide cold, with nitric acid, and then blow ethylene oxide through the solution. It reacted, $C_2H_4O + HBr \longrightarrow HOC_2H_4Br$, with the HBr to form ethylene bromohydrin, which was just blown out of the system. I considered this a neat trick.[*]

Anyway, both companies eventually came up with bids, and, as I suspected would be the case, Jefferson's was the better. They wanted fifteen dollars a pound for Cavea salt, while the best that Houdry could do, at that time, was seventy-five, although their research people thought that they might be able to argue their business end down to fifty. So now there was no supply problem to cope with. Howard Bost did some work with

which had a better freezing point than Cavea, and Charlie Tait phoned me on June 10 with some interesting news. Apparently Wyandotte had decided to give up the TNM monoprops as a lost cause, and to shift to the AN family. And they had some assorted substituted piperazines available, such as

And so Charlie bridged that with ethylene bromide, and came up with

The propellant had a freezing point well below −54°, and card-gapped at only three cards. Otherwise, it was just like Cavea. It was called Cavea B (the Rocket Branch thought that "2 methyl Cavea" would be too revealing a name!)

Wyandotte had made other similar compounds, some with two extra methyl groups, variously arranged, but Cavea B was the simplest and hence the best, and the others got nowhere. And Spurge Mobley had found himself a seven-membered piperazine-like compound

---

[*] He visited me some weeks later, and I asked him what Jefferson's substituted piperazines were used for. He answered, in a drawl as flat as Texas, "Well, they're a lot of farmers down our way, and they raise a lot of hawgs. And the hawgs get intestinal worms, and don't fatten up the way they should. So the farmer puts some of the piperazine in their feed, and the worm goes to sleep and forgets to hold on. And when he wakes up the hawg isn't there any more!"

and had bridged that one to produce the odd structure

$$
\begin{array}{c}
\text{C---C---C} \\
\text{C---N}^+\text{---C---C---N}^+\text{---C} \\
\text{C--------C}
\end{array}
$$

Spurge was highly indignant at me because, while his creation had the right freezing point, I noted that it was no improvement over Cavea B and cost several times as much, and so wouldn't push it.

Cavea B was the winner, and seemed to be the ideal monopropellant. And by the end of the year it had been fired successfully by NARTS, GE, Wyandotte, and Hughes Tool, with JPL soon to follow. It performed very well in a motor, yielding 94 or so percent of the theoretical impulse with a comparatively small chamber. Combustion was remarkably smooth—better than with the original Cavea (now called Cavea A) which was apparently just *too* symmetrical and stable to offer the combustion process any place to take hold. And there was no difficulty with supply. Wyandotte had any amount of the piperazine raw material.

Although the AN's were firmly planted in center stage at this time, other monopropellant systems were vigorously elbowing their way toward the spot light. For instance, Kenneth Aoki, at Wyandotte, made

$$
\text{O:N}\overset{\text{C---C}}{\underset{\text{C---C}}{\text{---C---C---}}}\text{N:O}
$$

the diamine oxide of triethylenediamine, and dissolved that in nitric acid. But he found that the heat of solution was so high (the acid probably decomposed the oxide and formed the nitrate instead) that any possible performance advantage over Cavea A or B was negated. He also made

$$
\text{O}_2\text{N---N}\overset{\text{CH}_2\text{---CH}_2}{\underset{\text{CH}_2\text{---CH}_2}{\text{---N---NO}_2}}
$$

intending to dissolve it in N_2O_4 or TNM, but found that it was too insoluble in those oxidizers to be of any use.

A more interesting approach to monopropellant development derived from the B–N propellant system, described in the boron chapter. As has been related, bipropellant B–N work had been plagued by combustion problems, and it was hypothesized—or hoped—that these might be alleviated if the boron and the nitrogen were combined in the same propellant—or even the same molecule.

McElroy and Hough, at Callery Chemical Co. started work on what they called the "Monocals." These consisted of an adduct, or addition compound of decaborane and two or three molecules of monomethyl hydrazine, all dissolved in about seven molecules of hydrazine. The mixing of the MMH with the decaborane had to be done in solution, or the product would explode when it warmed up. The propellants weren't particularly uniform, and varied considerably from lot to lot, for reasons not well understood, if at all. They were extremely viscous, but didn't appear to be particularly sensitive. Wyandotte tried them on the test stand, in a 50-thrust motor, with discouraging results. Bill Cuddy and his crew got off five runs altogether, four of them either starting or ending with a detonation and a wrecked motor. The Monocals died, unlamented, in 1960.

"Dekazine" lasted a little longer. In June of 1958, H. F. Hawthorne's group at Rohm and Haas prepared $B_{10}H_{12} \cdot 2NH_3$ by reacting bis (acetonitrile) decaborane with hydrazine. They had hoped to incorporate the N–N group of the hydrazine into the decaborane molecule, so as to get the closed cage carborane structure, but found that their product retained the open decaborane basket, and had no N-N bonds. At any rate, they dissolved one mole of it in about 7.5 of hydrazine (they couldn't get it to dissolve in fewer) and thought that they had a monopropellant. It wasn't the easiest thing in the world to live with. First, it picked up oxygen from the air. Second, it was thermally rather sensitive, starting to decompose exothermically at 127°. Its card-gap value was low—about 4 cards—but it was indecently sensitive to adiabatic compression, rating, on that test, between normal propyl nitrate and nitroglycerine. But they managed to get it fired, over the next couple of years, by Spencer King at Hughes and by Bob Ahlert at Rocketdyne at the 500-pound thrust level. Nobody ever got more than 75 percent of its theoretical performance out of it, and nobody could seem to find a way to prevent it from detonating (usually near the injector in a motor run) when it felt in the mood. Which was frequently—Ahlert survived some really impressive explosions. And so by the end of 1960, everybody gave up Dekazine as a bad job, and it was tenderly laid out on the marble slab next to that occupied by the late Monocal.

In 1959, however, Hawthorne of Rohm and Haas made an interesting observation, which was to lead to a different approach to the problem of a B-N monopropellant. He observed that when he reacted the bis (acetonitrile) (An) adduct of decaborane, in benzene at room temperature, with triethylamine (NEt_3) the reaction went mainly as below.

$$B_{10}H_{12}An_2 + 2NEt_3 \longrightarrow B_{10}H_{12}(NEt_3)_2 + 2An$$

However, if he ran the reaction at the boiling (refluxing) point of benzene, the reaction went almost quantitatively this way:

$$B_{10}H_{12}An_2 + 2NEt_3 \longrightarrow (HNEt_3^+)B_{10}H_{10}^= + 2An$$

The open decaborane basket had closed up to the 10-cornered, 16-faced closed-cage structure of what was later called the "perhydrodecaborate ion." This was a remarkably stable structure, the anion of a very strong acid—almost as strong as sulfuric acid. It was no trick to get the hydrazine salt of this acid (several simple routes were discovered that same year), $(N_2H_5)_2B_{10}H_{10}$. The unsolvated salt was shock-sensitive, but when crystallized with either one or two molecules of hydrazine—it was easy to get either form—it was safe and easy to handle. And it could be dissolved in hydrazine to make a propellant.*

Unfortunately, you couldn't dissolve enough of it in hydrazine to get the number of B atoms in the mixture to equal the number of N atoms, which is what you want with a B–N monopropellant. Lou Rapp, at Aerojet (he had recently moved there from Reaction Motors) somewhere around the beginning of 1961, thought that if he substituted one of the "Hepcat" cations for the hydrazinium ions in the salt, he might be able to remedy that deficiency, I had the same idea at the same time—and I could move faster than he could. My outfit was small, and I didn't have to pay any attention to contracts, since I didn't have any, the brass seldom paid any attention to what I was doing, and I could usually try whatever I wanted to try before anybody in authority could get around to telling me not to. So it went like a breeze.

I had Mobley make up a few grams of the Hepcat chloride by milling ammonium chloride and lithium borohydride together, as has already been described. And I had him make up some potassium perhydrodecaborate. Then he dissolved the two in liquid ammonia, and mixed them together in the proper proportions. The reaction went

$$K_2B_{10}H_{10} + 2[BH_2(NH_3)_2]Cl \longrightarrow$$

$$[BH_2(NH_3)_2]_2B_{10}H_{10} + \underline{2KCl.}$$

The potassium chloride precipitated, and was filtered off, the ammonia was allowed to evaporate, and I had the Hepcat perhydrodecaborate. After making sure that I had what I thought I had—using IR and so on for diagnosis—I had him add one mole of it to six moles of hydrazine. Four of the hydrazines displaced the ammonias in the cations and it bubbled off, and two were left over as solvent. So I finally had $[BH_2(N_2H_4)_2]_2B_{10}H_{10} + 2N_2H_4$. Here I had two borons in the cations, and ten in the anion. There were eight nitrogens in

---

* I had the bright idea that it might also be used to make a peroxide-based monopropellant. I had some of the ammonium salt of the perhydrodecaborate ion made, and put a few milligrams of it on a watch glass. I then put a drop of concentrated H_2O_2 next to the salt, and tilted the watch glass to bring the two into contact. There was a brilliant green-white flash, and the sharpest detonation I have ever heard. The watch glass was reduced, literally, to a fine powder. End of bright idea.

the cations, and four in the solvent. And finally, there were twenty hydrogens in the cations, ten in the anion, and eight in the solvent, so that the whole mess balanced to $12BN + 19H_2$. And miraculously, the thing was liquid at room temperature, and not too viscous, and didn't appear to be particularly sensitive. We had only a few cc of the stuff, but it looked interesting.

The psychological payoff came at a monopropellant meeting in August. Lou Rapp described what he was trying to do, and I then took a sadistic delight (I was chairman of that session) in pulling the rug out from under him by pointing out that we had already done it, and describing how. Lou Rapp and I were good friends, but it wasn't often that I had a chance to do him in the eye, and it was too good an opportunity to miss.

However, that was about as far as that propellant ever got. The Army brass (the Navy had moved out a year before, and the Army had taken over NARTS, which became the Liquid Rocket Propulsion Laboratory of Picatinny Arsenal) passed down the word that the Army had no interest in BN monopropellants, and to knock it off. I believe that their decision was the right one. My monster would have been horribly expensive to make, its density was by no means impressive, and there was no a priori reason to believe that it would perform any better than the other B–N monoprops. The stuff was not a practicable propellant. The whole performance had been a tour de force designed to show that a balanced B–N monoprop *could* be made. It was a lot of fun, but it was the end of the B–N monopropellants.

All this interest in monopropellants had led to the formation, at the first monopropellant conference in November 1953, of the Monopropellant Test Methods Committee, which was operated first under the sponsorship of BuAer, then under the American Rocket Society, and then under Wright Air Development Center. In November 1958 its field was extended to cover all liquid propellants, and the Liquid Propellant Information Agency took it over, and it's still in operation. I served on it, on and off, for several years.

The original reason for its formation was the inherent instability of monopropellants. *Any* monopropellant with a reasonable amount of energy in it can be detonated if you go about it the right way. Everybody in the business had his own pet method of measuring the sensitivity of the monoprops he was working with. The only difficulty was that no two methods were alike, and it was just simply impossible to compare the results from one laboratory with those from another. In fact, it was just about impossible to define, say, shock sensitivity, at all. Even the *relative* sensitivities of two propellants might depend—and often did—upon the apparatus used to make the measurement. The job of the committee was to examine all the methods used, to pick out those which gave more or less reproducible results, or to talk people into developing such methods, then to standardize these, and finally to try to persuade the people in the field to *use those methods*. Then, it was hoped, even if none of our tests or results made sense we would be unanimous in our fantasies and could talk to each other with something approaching coherence, and with luck, comprehension by the hearer.

The first test that was adopted in July 1955 was the card-gap test, previously described. It had first been developed by the Waltham Abbey People, and then streamlined by NOL, and we spent a lot of time chasing down anomalous results. I remember Joe Herrickes, of the Bureau of Mines, and I firing our two card-gap setups side by side for a whole day, trying to discover why, with one monopropellant they agreed beautifully, and with another disagreed wildly. We finally discovered, after interchanging practically everything in the setups, that the trouble lay in the sample cups. Joe's were made of aluminum, mine of iron pipe. The committee standardized on the iron pipe, and we published. Putting a final report together was sometimes something of a Donnybrook. The committee, as might be imagined, was composed of highly self-confident and howlingly articulate individualists, and there were always at least six of us present each of whom considered himself a master of English prose style. Whew!

A test that took a lot longer to decide on was the "Drop-Weight." For years, people in the explosive business had been dropping weights on samples of their wares, and rating their sensitivity on the basis of how far you had to drop how big a weight in order to make the sample go off. We looked into the matter and discovered, to our dismay, that the JPL tester disagreed with the Picatinny tester, which disagreed with the Hercules apparatus, whose results could not be compared with those of the Bureau of American Railroads, which, in turn, contradicted those of the Bureau of Mines. Furthermore, none of them was any good at all with liquids.

Bill Cuddy, of Wyandotte, in March 1957 described a tester designed specifically for liquids; it was modified by Don Griffin of Olin Mathieson and finally evolved into what is known as the OM drop weight tester.[*] Actually it was a device for measuring adiabatic sensitivity. The falling weight suddenly compressed a tiny and standardized air bubble above the one drop sample, and the adiabatic heating of this bubble set the thing off—or wasn't enough to do it, as the case might be. It was, and is, quite a satisfactory instrument once you got used to its little foibles. For instance, it has to be on a *really* solid foundation if you hope to get reproducible results. We ended up with the instrument bolted to a three-foot square of three-inch armor plate, which was in turn bolted to a six-foot cube of concrete which rested on bedrock—granite. That way, it worked fine.

Al Mead, of Air Reduction, came up in 1958 with the standard thermal stability tester, another very useful instrument. In it, a small sample was heated at a constant rate, and the temperature at which the sample started to warm up

---

[*] Don Griffin, a free soul if I ever knew one, then took a year's vacation from rocket propulsion, spending it in the Hula-Hoop business. He said it made more sense.

faster than the heating bath was taken as the takeoff point. We used it, along with our own, for years. They really measured different things, and we could use them both.

Other tests were standardized and published, but these were the more useful and most often used. One subject that was investigated for years was detonation velocity, the critical diameter for detonation, and the construction of detonation traps. If you have a monopropellant blow in your motor, that's one thing. But if that detonation propagates back (at some 7000 meters per second, usually) through the propellant line to the propellant tank, and *that* blows, then you can be in real trouble. If the diameter of the propellant line is small enough, the detonation will not propagate and dies out—the limiting diameter being called the "critical diameter." It varies with the nature of the propellant, the material of which the line is made (steel, aluminum, glass, etc.) with the temperature, and maybe with a few more things. (When we found that a detonation in Isolde would propagate nicely through hypodermic-needle tubing, our hair stood on end, and we perspired gently.)

Starting around 1958, and carrying on into 1962, a lot of work on detonation propagation and trapping was carried on at Rocketdyne, Wyandotte, JPL, BuMines, GE, Hughes, Reaction Motors, and NARTS-LRPL, and valiant efforts—some of them successful—were made to find a way to stop a detonation in its tracks. Because *that* can be done, sometimes, by putting a detonation trap in the line. Designing such a thing is not a scientific matter; it's a piece of empirical engineering art. And the various designs showed it. One early trap consisted simply of a loop in the line—like so: ⌒ When the detonation came barreling down the line, blowing up the tubing as it came, it would cut the other part of the line where they crossed, leaving the detonation no place to go. This wasn't too reliable. The bang that cut the line might start another, brand new, detonation going. Bob Ahlert did very well by putting a section of Flex-Hose, a Teflon tube reinforced with a metal wire mesh, in his line. The detonation would simply blow this weak section out, and then have nowhere to go. And Mike Walsh, of our group, devised a trap that worked beautifully with Cavea B, as well as with several other monopropellants. Cavea B would not propagate a detonation through a 0.25-inch line, but would through a one-inch line. So Mike inserted a one-foot long piece of two-inch piping into his one-inch line, and filled this section with a cylindrical bar of a plastic that would resist the propellant for a reasonable time. (Polystyrene was good.) And he drilled sixteen 0.25-inch holes longitudinally through this cylindrical plug, so that he had the same area for flow as he had in the one-inch main line. And when he checked it out, the detonation rolled down to the trap, blew up about the first third of that, and stopped cold.

But detonation traps aren't always the complete answer. We discovered that when, in the summer of 1960, we tried to fire a 10,000-pound thrust Cavea B motor. We didn't have Mike's trap at that time, so we inserted a battery of six-teen 0.25-inch loop traps in the line. Well, through a combination of this and that, the motor blew on startup. We never discovered whether or not the traps worked—we couldn't find enough fragments to find out. The fragments from the injector just short-circuited the traps, smashed into the tank, and set off the 200 pounds of propellant in that. (Each pound of propellant had more available energy than two pounds of TNT.) I never saw such a mess. The walls of the test cell—two feet of concrete—went out, and the roof came in. The motor itself—a heavy, workhorse job of solid copper—went about 600 feet down range. And a six-foot square of armor plate sailed into the woods, cut-ting off a few trees at the root, smashing a granite boulder, bouncing into the air and slicing off a few treetops, and finally coming to rest some 1400 feet from where it started. The woods looked as though a stampeding herd of wild elephants had been through.[*]

As may be imagined, this incident tended to give monopropellants some-thing of a bad name. Even if you could fire them safely—and we soon saw what had gone wrong with the ignition process—how could you use them in the field? Here you have a rocket set up on the launching stand, under battlefield conditions; and what happens if it gets hit by a piece of shrapnel? LRPL came up with the answer to that. You keep your monoprop in the missile in two com-partments: one full of fuel-rich propellant made up to $\lambda = 2.2$ or 2.4, and the other containing enough acid to dilute it to $\lambda = 1.2$. Just before you fire, a can-opener arrangement inside the missile slits open the barrier separating the two liquids, you allow a few seconds for them to mix, and then push the button. The idea—it was called the "quick mix" concept—worked fine. We couldn't use a double-ended compound like Cavea A or B for the propellant—made up to 2.4 the freezing point was too high—so we first tried "Isobel," the dimethyldiisopro-pyl ammonium nitrate, I phoned a chemical company in Newark to have them make us a hundred pounds or so of the salt, and was assured by their director of research that it couldn't be done, since the compound was sterically impos-sible. I protested that I was staring at that moment at a bottle of the salt sitting on my desk in front of me, but I couldn't convince him. Isobel, however, didn't quite meet the freezing point limitations, so we shifted to Isobel E and Isobel F, the diethyldipropyl, and the ethyltripropyl salts, respectively, which did. These

---

[*] When I got down to the test area after the bang, one of the rocket mechanics galloped up and demanded, "My God, Doc! What the Hell did you send us this time?" The only response of which I was capable was to light a cigarette and remark, "Now, really, Johnny! You should see my Martinis!" What got me though, was the remark by an officer from Picatinny, after viewing the mess. This was just before NARTS was due to be "disestablished" and taken over by the Army, and this character (metaphorically) held his nose and stated to nobody in particular, "*Now* I know what the Navy means by 'disestablishment.'" I wanted to kill him.

didn't have quite the thermal stability I wanted (usually the higher the λ the worse the stability) so I finally came up with Isobel Z, diethyldiisopropyl ammonium nitrate, which was immensely better. (If that character in Newark thought that Isobel was impossible, I wonder what he would have thought of Isobel Z! There actually isn't any trace of steric hindrance in the ion, but it's a near thing. Not one more carbon atom could have been crammed into the space around the nitrogen.) LRPL tried the fuel-rich Isobels, at 2.4 or so, as low-energy monopropellants, for APU work and so on, and found that they worked very well in such a role. That work was pretty well cleaned up by 1962.

Dave Gardner, of Pennsalt, didn't anticipate any detonation problems when he started work in May of 1958. What the Air Force wanted from him was a low-energy monoprop for APU work, and one with such thermal stability that it could survive at 200–300°C, and Dave naturally figured that a low-energy compound with that sort of stability wasn't going to give him much trouble. He worked for about three years on the job—Sunstrand Machine Tool did his motor work—at first with mixtures of extremely stable—and hence low-energy—fuels in various acid oxidizers. His fuels were sometimes salts; the tetramethyl ammonium salt of sulfuric acid, or of fluorosulfonic acid, or of trifluoromethane sulfonic acid, or the pyridine salt of sulfuric acid. Sometimes he used methane or ethane sulfonic acid itself as a fuel, or the fluorinated $CF_3CH_2SO_3H$. His oxidizers were either perchloric acid dihydrate, or a mixture of nitrosylpyrosulfate and sulfuric acid, sometimes with some nitric acid added. Performance—which he wasn't looking for—was naturally poor, but he had some remarkably heat-resistant propellants on his hands.

A little later, impressed by the remarkable stability of the—$CF(NO_2)_2$ group he synthesized the monopropellant $CH_3CF(NO_2)_2$, which he called "Daphne." (I never discovered what the name celebrated—or commemorated.) The performance wasn't particularly startling—that C-F bond is awfully strong—but the material appeared to be stable to just about everything. But, alas, Daphne, too, was a woman, and could betray him. And took out most of the test cell when she did.

Incidental to his monopropellant work, Dave produced a pair of rather interesting high-density oxidizers, by fluorinating potassium nitroformate, $KC(NO_2)_3$ and similar compounds. They were $F—C(NO_2)_3$ which he called

D-11, and
$$\underset{\displaystyle NO_2}{\overset{\displaystyle NO_2}{F—C}}—\underset{\displaystyle NO_2}{\overset{\displaystyle NO_2}{C}}—F$$
called D-112. He measured their heat of

formation by reacting them in a bomb calorimeter with carbon monoxide, and when he reported his results insisted on putting my name on the paper as coauthor (which I was not) because I'd tried that reaction first, with tetranitromethane, and had suggested it to him as a good way to get a handle on the thermochemistry of highly oxygenated compounds. The oxidizers appeared, unfortunately, just in time to be made obsolete by ClF_5.

One of the strangest, and certainly the hairiest, chapter in the propellant story had its start in two unrelated but almost simultaneous events. First, the Advanced Research Projects Administration (ARPA) initiated "Project Principia" early in June 1958. The idea was to get certain large chemical companies into the business of developing improved solid propellants. "To get the benefit of their fresh thinking," was the phrase. American Cyanamid, Dow Chemical, Esso R and D, Minnesota Mining and Minerals (3M), and Imperial Chemical Industries were those invited to enter the field.

The old line propellant people—Reaction Motors, Aerojet, Rocketdyne—not unnaturally raised a row. In fact, they screamed from the rooftops. "These newcomers may have all sorts of fancy equipment, and they may be hot on molecular orbital theory and pi bonds, but they don't know a propellant from a panty raid, and what you'll get is a lot of expensive and useless chemical curiosities. *We,* on the other hand, have been living with propellants for years, and know what they have to do, and we're not such bad chemists ourselves, and just get us the same fancy equipment and anything they can do we can do better!" But their protests fell on unheeding ears, and "Principia" got under way. (In the end, the professional propellant people turned out to be pretty good prophets.)

The second event was the announcement by W. D. Niederhauser of Rohm and Haas, at a propellant meeting late in September, that N_2F_4, in a gas phase reaction, would add across an olefinic bond, forming the structure

$$\begin{array}{ccc} H & & H \\ | & & | \\ -C & - & C- \\ | & & | \\ NF_2 & & NF_2 \end{array}$$ and that this was a general reaction, applicable to practically

any olefin. He also announced that HNF_2 could be prepared by reacting N_2F_4 with AsH_3.

So all the companies concerned in Principia started reacting N_2F_4 with various olefins to see what would happen. (Frequently what happened was that the reactor blew up.) Rohm and Haas was, of course, already doing this sort of thing, and at the beginning of 1959 both Reaction Motors and Jack Gould at Stauffer dropped the monopropellant systems they had been investigating and joined the new field of "N-F" chemistry. And Callery Chemical Co. quickly did the same. So there were soon at least nine organizations involved in this sort of research.

A few glitches interfered with the march of science, and the progress of the solid propellant program. The first was that practically none of the new compounds discovered were solids. Most of them were liquids, and rather volatile liquids at that. ARPA bowed to fate and the facts of life, and in November of 1960 rewrote the contracts to include work on liquids. The second glitch could not, unfortunately, be cured by making a mark on a piece of paper.

Practically every compound discovered turned out to be indecently sensitive and violently explosive. For instance, take the case of the simplest NF compound, the first one discovered. Thermodynamically, it tends to react

$$\begin{array}{ccc} H & H \\ | & | \\ H-C-C-H \\ | & | \\ NF_2 & NF_2 \end{array} \rightarrow 4HF + 2C.$$ And given the slightest provocation, it

will do just that. And the formation of four molecules of HF produces enough energy to make the resulting explosion really interesting. The things were shock sensitive. Many of them were heat sensitive. And some were so spark sensitive that they couldn't be poured. When you tried it, the tiny static charge formed was frequently enough to trigger a detonation. The only way to reduce their sensitivity, apparently, was to make a compound with a large molecule and a small number of NF_2 groups. Which was not good for its energy content.

Another thing that was bad for their (prospective, and eventually, maybe) propellant performance was the fact that on decomposition they produced considerable quantities of free carbon, which, as has been explained, is not good for performance. The obvious thing to do was to incorporate enough oxygen in the molecule to burn the carbon to CO. Dow, in the spring of 1960, synthesized such a compound, balanced to CO by the direct fluorination of

urea. Its structure was $$F_2N-\overset{\overset{\textstyle O}{\|}}{C}-NH_2$$ and it is perhaps just as well that it was produced in very small quantities, as it was indecently spark sensitive. Jack Gould, in 1961, reported on F_2N-CH_2-OH and it was synthesized by I. J. Solomon of the Armour Research Institute, early in 1963, by the mole for mole reaction of HNF_2 and formaldehyde. Again—impossibly sensitive. Allied Chemical Co., in 1962, and the Peninsular Chemical Co., about four years ago, tried to synthesize its isomer CH_3-O-NH_2. If they had succeeded, it would have served them right. I know from my own experience that methoxy amine, CH_3-O-NH_2, which can decompose at worst and most energetically, only to CO and NH_3 plus a little hydrogen, is quite sensitive on the OM drop-weight apparatus. What a similar compound, whose decomposition would lead to the formation of two HF molecules would be likely to do, simply boggles the mind.

Esso synthesized a number of nitrate and nitramine derivatives of NF compounds, of which $$\begin{array}{c} NF_2 \\ | \\ O_3N-CH_2-C-CH_2-NO_3 \\ | \\ NF_2 \end{array}$$ first described early in 1962, is typical. Some few of the series were solids rather than liquids, but all were impossibly sensitive. This compound is interesting, however, in having two NF_2 groups on the same carbon atom. This structure was made possible by a reaction discovered by the Rohm and Haas people in 1960–61. If HNF_2 is reacted with an aldehyde or ketone in concentrated sulfuric acid, the reaction goes

$$2HNF_2 + R\text{---}\overset{\overset{\textstyle O}{\|}}{C}\text{---}R \rightarrow R\text{---}\underset{\underset{\textstyle NF_2}{|}}{\overset{\overset{\textstyle NF_2}{|}}{C}}\text{---}R + H_2O$$ and you are left with

two NF_2 groups on the same carbon. The reaction was quite general, and a wide variety of "geminal" difluoramino compounds were synthesized. They were just as sensitive as the "vicinal" or $\text{---}\underset{\underset{\textstyle NF_2}{|}}{\overset{\overset{\textstyle H}{|}}{C}}\text{---}\underset{\underset{\textstyle NF_2}{|}}{\overset{\overset{\textstyle H}{|}}{C}}\text{---}$ type.

Another way to get oxygen into the monopropellant was to mix the NF compound with an oxygen-type oxidizer. Jack Gould (Stauffer) came up in 1961 with a concoction he called "Hyena," which consisted of an NF (usually $F_2NC_2H_4NF_2$) dissolved in nitric acid. J. P. Cherenko, of Callery, produced similar mixtures (called "Cyclops" this time) but he sometimes used N_2O_4 or tetranitromethane instead of the acid, and sometimes tranquilized the propellant (he hoped) by adding pentane. Hyena and Cyclops were both unmitigated disasters. The man who was determined to make an NF monopropellant work, or to prove, definitely, that it couldn't be done, was Walt Wharton of the Army Missile Command, at Huntsville, and from the middle of 1961 to the end of 1964 he and Joe Connaughton valiantly and stubbornly pursued that objective. His chosen compound was IBA, the IsoButylene Adduct of N_2F_4, made by

Rohm and Haas by the reaction $CH_3\text{---}\overset{\overset{\textstyle CH_2}{\|}}{C}\text{---}CH_3 + N_2F_4 \rightarrow CH_3\text{---}\underset{\underset{\textstyle NF_2}{|}}{\overset{\overset{\textstyle CH_2NF_2}{|}}{C}}\text{---}CH_3.$

If the compound is mixed with N_2O_4 (1.5 molecules of the latter to one of IBA) the mixture is a monopropellant with a good density and a fairly attractive (theoretical) performance—293 seconds. Other compounds containing more NF_2 groups would have given more, but the idea was to get *any* NF to work at all.

IBA, straight, was extremely sensitive on the OM drop-weight apparatus, and Wharton was immensely encouraged, at first, to discover that the addition of a very small amount of N_2O_4—less than 1 percent—cut this sensitivity down to practically nothing. But then he started burning-rate studies, in a liquid strand burner. He had ignition problems—a hot wire wasn't too reliable—and he discovered that the burning rate of the material was vastly sensitive to the bomb pressure. (A trace of ferric chloride would decrease the rate, and one of carbon tetrachloride would increase it.) He furthermore discovered that the material had a lamentable tendency to detonate in the bomb or in a motor, that a glass tube detonation trap wasn't particularly helpful, and he was made a bit thoughtful by the discovery that the critical diameter for detonation propagation was less than 0.25 millimeters—less than 0.01 inch.

Most of his early work was done with a "T" motor, little more than an injector and a nozzleless chamber, with observation ports so that the ignition process could be observed with a high-speed camera. They tried a slug of ClF_3 for ignition, and got a detonation instead. They eventually settled on a slug of antimony pentachloride—of all things—which gave a smooth and reliable start. By this time they were working with an "expendable" motor with no nozzle and a low chamber pressure, about two atmospheres, and were mixing the N_2O_4 and IBA remotely right on the test stand. It was fortunate that they worked remotely, since 150 cc of the mixture detonated during a run and wrecked the setup.

In the winter of 1962–63 they sent a sample of IBA (dissolved in acetone so it could be transported more or less safely) up to LRPL for card-gap work. We gently distilled off the acetone, and made the tests. (Mixing the IBA and the N_2O_4 was a precarious business.) Straight IBA wasn't particularly sensitive on card-gap, about ten cards, and the material with 1 percent of N_2O_4 in it was about the same. But when mixed up for maximum performance—one mole of IBA to 1.5 of N_2O_4—the sensitivity was more than 96 cards. We never discovered how much more; our interest in the subject had evaporated.[*]

---

[*] Two people can operate the card-gap apparatus, and three operators is optimum. But when LRPL did this particular job (the feather-bedding at Picatinny was outrageous) there were about seven people on the site—two or three engineers, and any number of rocket mechanics dressed (for no particular reason) in acid-proof safety garments. So there was a large audience for the subsequent events. The old destroyer gun turret which housed our card-gap setup had become a bit frayed and tattered from the shrapnel it had contained. (The plating on a destroyer is usually thick enough to keep out the water and the smaller fish.) So we had installed an inner layer of armor plate, standing off about an inch and a half from the original plating. And, as the setup hadn't been used for several months, a large colony of bats—yes, bats, little Dracula types—had moved into the gap to spend the winter. And when the first shot went off, they all came boiling out with their sonar gear fouled up, shaking their heads and pounding their ears. They chose one rocket mechanic—as it happens, a remarkably goosy character anyway—and decided that it was all his fault. And if you, gentle reader, have never seen a nervous rocket mechanic, complete with monkey suit, being buzzed by nine thousand demented bats and trying to beat them off with a shovel, there is something missing from your experience.

These figures did not encourage further work with the IBA-N_2O_4 mixture. There was some talk of using the combination as a bi-propellant, but that would have been rather pointless. Wharton and Connaughton fired the straight IBA as a monopropellant, at the 250-pound thrust level, but there was so much free carbon in the exhaust that they never got more than 80 percent even of its rather low specific impulse. They were driven, reluctantly, to the conclusion that an NF monopropellant was not practical politics, and abandoned the whole idea late in 1964.

Just as Wharton was starting his IBA work, there occurred one of the weirdest episodes in the history of rocket chemistry A. W. Hawkins and R. W. Summers of Du Pont had an idea. This was to get a computer, and to feed into it all known bond energies, as well as a program for calculating specific impulse. The machine would then juggle structural formulae until it had come up with the structure of a monopropellant with a specific impulse of well over 300 seconds. It would then print this out and sit back, with its hands folded over its console, to await a Nobel prize.

The Air Force has always had more money than sales resistance, and they bought a one-year program (probably for something in the order of a hundred or a hundred and fifty thousand dollars) and in June of 1961 Hawkins and Summers punched the "start" button and the machine started to shuffle IBM cards. And to print out structures that looked like road maps of a disaster area, since if the compounds depicted could even have been synthesized, they would have, infallibly, detonated instantly and violently. The machine's prize contribution to the cause of science was the structure, $H\text{---}C\equiv C\text{---}N\text{------}N\text{------}H$ to

$$\begin{array}{cc} | & | \\ O & O \\ | & | \\ F & F \end{array}$$

which it confidently attributed a specific impulse of 363.7 seconds, precisely to the tenth of a second, yet. The Air Force, appalled, cut the program off after a year, belatedly realizing that they could have got the same structure from any experienced propellant man (me, for instance) during half an hour's conversation, and at a total cost of five dollars or so. (For drinks. I would have been afraid even to *draw* the structure without at least five Martinis under my belt.)

The NF programs led to some interesting, if eventually unproductive, oxidizer work. It was obvious, very early in the game, that if you could tie enough NF_2 groups to a carbon atom, the result would be more a fluorine-type oxidizer than a monopropellant. Cyanamid, late in 1959, took the first step in this direction when they synthesized $F_2N\text{---}C\text{==}NF$ Then 3M, in the spring of

$$| \\ F$$

1960, synthesized "Compound M," $F_2C(NF_2)_2$ by the direct fluorination of ammeline, and a little later came up with "Compound R," $FC(NF_2)_3$ by the same route. Dow and 3M, in 1960, both synthesized perfluoroguanidine, or "PFG" $FN=C(NF_2)_2$ by the reaction of fluorine diluted with nitrogen on guanidine. And finally, in 1963, "Compound Δ" (delta) or "T" or "Tetrakis"— from tetrakis (difluoramino)methane—$C(NF_2)_4$ was synthesized at Cyanamid by Frank, Firth, and Myers, and by Zollinger at 3M. The former had fluorinated the NH_3 adduct of PFG, the latter had used the HOCN adduct.

All these compounds were difficult to make—only R ever achieved synthesis in pound lots—and incredibly expensive. Their calculated performances, with suitable fuels, was impressive enough, but their sensitivity was even more so. None of them could be lived with. Attempts *were* made to tranquilize them by mixing them with less temperamental oxidizers, but the results were not happy. Wharton worked for some time with a mixture of R and N_2O_4, and Aerojet tried some mixtures (called "Moxy"), comprising R, N_2F_4 and ClO_3F, or Δ, N_2F_4 and ClO_3F. But it was hopeless. When the NF oxidizer was sufficiently diluted to be safe, all its performance advantage had gone with the wind.

There was some thought that an OF structure attached to the carbon would be more stable than the NF_2 structure; and in 1963 W. C. Solomon of 3M showed the way to such structures by reacting fluorine with oxalates suspended in perfluorokerosene, in the presence of a transition metal, to get $F_2C(OF_2)_2$. Three years later, Professor George Cady's group, at the University of Washington, synthesized the same compound, neatly and elegantly, by reacting fluorine and carbon dioxide, at room temperature, in the presence of cesium fluoride. But the very mildness of the conditions for its synthesis showed that it was too stable to be of much use as an oxidizer. And, finally, as has been mentioned in the halogen chapter, the group at Allied Chemical, reacted ONF_3 with a perfluoro olefin, such as tetrafluoroethylene, to get $CF_3—CF_2—ONF_2$ or one of its cousins. But an ONF_2 group attached to a heavy and remarkably stable fluorocarbon residue isn't very useful in the rocket business.

So in the long run, NF programs didn't lead to much in the way of practical liquid propellants, brilliant as was some of the chemistry exhibited. The record of this chemistry is now being collected, to be embalmed safely in a definitive text, so that nobody will ever, ever, have to risk his neck doing it again.

As for the original object of Principia: solid propellant grains containing NF_2 groups *have* been made—and fired. But they have a long way to go and if they are operational before 1980 or so I, for one, will be surprised.

And as for the future of the high-energy monopropellants: I'm afraid that it's in the past. We all worked for years trying to reconcile properties which we finally and sadly concluded were irreconcilable—high energy and stability. For all our efforts, no high energy monoprop has made the grade to operational status. Cavea B almost made it, but "almost" is not success. But it was a damned good try!

12

High Density
and the Higher
Foolishness

The idea of a hybrid rocket, one using a solid fuel and a liquid oxidizer is a very old one; in fact, Oberth had tried to make one for UFA back in 1929, and BMW had experimented with such a device during 1944–45. Configurations vary somewhat, but the usual arrangement is a cylinder of fuel, solid except for a longitudinal passage down the center line, fitted tightly into a cylindrical chamber. Oxidizer is injected at the upstream end, and reacts with the fuel as it travels down the passage, and the combustion products eventually emerge through the nozzle just downstream of the fuel grain. (Even if it weighs two hundred pounds, it's still a "grain.")

On the face of it, the idea looks attractive. Solid fuels are denser than liquid fuels, for one thing, and for another, the rocket can be throttled just like a pure liquid device, while there is only one liquid to handle. From the point of view of safety, it looks ideal, since there just isn't any way for the fuel and the oxidizer to get together until you want them to.

Soon after the end of the war, then, several organizations set out confidently to design—and fire—hybrid rockets, and fell flat on their corporate faces. The experience of GE (in 1952, on Project Hermes) was typical. Their intention was to use a polyethylene fuel grain, with hydrogen peroxide as the oxidizer. And when they fired their rocket, the results were worse than depressing—they were disastrous. Combustion was extremely poor, with a measured C^* to make an engineer weep. And when they tried to throttle their motor, the oxidizer-fuel ration varied madly, and was never anywhere near the optimum for performance. (This is hardly surprising, since the oxidizer consumption depends upon the rate at which it is injected, while the fuel consumption depends on the area of the fuel grain exposed.) And tinkering with the injector and the exact shape of the fuel grain did very little good.

The engineers had been guilty of a sin to which engineers are prone— starting their engineering before doing their research. For it had become devastatingly clear that *nobody* knew *how* a solid fuel burned. Did it evaporate, and then burn in the vapor phase? Or was a solid-state reaction involved? Or what? There were lots of questions, and very few answers, and hybrid work languished for some years. Only the Navy, at NOTS, kept at it, trying to learn some of the answers.

The revival started in 1959 when Lockheed, with an Army contract, started hybrid work. In 1961 ARPA got into hybrids in a big way, and by 1963 there were at least seven hybrid programs going.

I was greatly amused by the behavior of each new contractor as it got into the act. The pattern was invariable. First, they would get themselves a computer. Then, they would calculate the performance of every conceivable liquid oxidizer with every conceivable solid fuel. And then they would publish a huge report containing all the results of all of these computations. And to the surprise of nobody who had been in the business any length of time (we had all made these calculations for ourselves years before) everybody came out with the same numbers and recommended practically identical combinations. Thus, the fuel grains recommended by three different contractors, Lockheed, United Technology Co., and Aerojet, comprised:

1 Lithium hydride plus a hydrocarbon (rubber) binder;
2 Lithium hydride plus lithium metal plus a binder;
3 Lithium hydride plus powdered aluminum plus a binder.

And the oxidizers recommended (not necessarily in the same order) consisted of:

1 Chlorine trifluoride plus perchloryl fluoride;
2 The same two plus bromine pentafluoride;
3 Or, plus N_2F_4;
4 Or, finally, and a little further out, straight OF_2.

All of which made some of us wonder whether or not the taxpayer had got his money's worth from all that expensive computer time.

Rohm and Haas investigated an entirely different type of hybrid, one which would still burn and produce thrust even when the oxidizer was completely cut off. The grain consisted of aluminum powder, ammonium perchlorate, and a plastisol binder. (Plastisol is a castable and quick-curing double-base mixture, consisting largely of nitrocellulose and nitroglycerine, and was a solid propellant in its own right.) Its combustion products included a large fraction of hydrogen and carbon monoxide, and the liquid oxidizer, N_2O_4, was intended to react with this and to increase the energy output and the thrust. NOTS performed a long series of combustion studies with a similar system, RFNA oxidizer and a fuel rich composite grain (ammonium perchlorate and a hydrocarbon or similar binder). As the hybrid system is a compromise between a solid and a liquid system, these and similar combinations can be considered as combining hybrid and solid features.

Steve Tunkel at Reaction Motors investigated a much more esoteric system in 1962–63—a reverse hybrid in which the oxidizer was in the grain, which consisted of nitronium perchlorate, NO_2ClO_4 or hydrazine di-perchlorate, $N_2H_6(ClO_4)_2$, in a fluorocarbon (Teflon-type) binder. The liquid fuel was hydrazine, and powdered aluminum or boron could either be suspended in the fuel or incorporated into the grain. The idea was to let the fluorine in the fluorocarbon react to form aluminum or boron trifluoride, while the carbon was oxidized to CO. (The other combustion products would depend upon the exact grain composition, the fuel flow, and so on.) The idea was interesting, but their hopes were never realized. Nitronium perchlorate turned out to be inherently unstable, for one thing, and Tunkel was never able to achieve efficient fluorocarbon-metal combustion. The system was just too precious to work.

Much more important, in the long run, was some of the work at UTC, who had a Navy contract to investigate the basic mechanism of hybrid combustion. (This, of course, should have been done at least ten years earlier, and before a lot of money had been sunk into hybrid work. But it's always easier to get money for engineering than for fundamental research. Don't ask me why.)

Most of this work was done with a simplified model of a hybrid motor, consisting of a flat slab of fuel with the oxidizer flowing across its surface, the whole in a transparent chamber so that the investigators could see what was happening, and take pictures of it. The fuel was usually polyethylene or methyl methacrylate (Plexiglass) and the oxidizer was oxygen or OF_2. They learned that the oxidizer reacts with the fuel only in the vapor phase, and that the rate was controlled by diffusion, while the rate of regression (consumption) of the fuel depended largely upon heat transfer from the hot reacting gases. (This, of course, was not strictly true when the fuel grain contained oxidizer of its own.) They learned that proper injector design could keep the regression rate uniform across the whole grain surface, but that the mixing of the fuel vapor and the oxidizer was so slow that additional mixing volume downstream of the grain was usually necessary to get reasonable combustion efficiency. This extra volume did much to reduce the density advantage claimed for the hybrid systems. But they learned how to build a hybrid motor that would work with reasonable efficiency.

Thus, although all the work with the lithium hydride grains and the chlorine trifluoride oxidizers never led to anything in particular, the fundamental research done at UTC led eventually to one hybrid motor which is operational and flying—the UTC power plant of a target drone. The oxidizer is N_2O_4, and the fuel is a very fuel-rich composite solid propellant. A hybrid motor could be made and made to work—but the hybrid was not the answer to everything, and its place in the propulsion spectrum is, and will be, very limited.

The "Arcogels" were another attempted approach to a high-density system. These were conceived in 1956 by the Atlantic Research Co., who worked on them for some five years. They were a mixture comprising mainly powdered ammonium perchlorate, aluminum, and a relatively nonvolatile liquid fuel and carrier, such as dibutyl phthalate. They had the consistency, approximately, of toothpaste. They obviously couldn't be brought into a chamber through a normal injector, but had to be forced in through special burner tips, which spread the pasty ribbon out to expose the maximum burning area. They burned all right, at least on a small scale, but their high density wasn't enough of an advantage to outweigh the horrendous problem of designing an injection system that could be carried in a flyable missile, and they never got anywhere.

All sorts of efforts were being made, during the late 50's, to increase propellant densities, and I was responsible (not purposely, but from being taken seriously when I didn't expect to be) for one of the strangest. Phil Pomerantz, of BuWeps, wanted me to try dimethyl mercury, $Hg(CH_3)_2$, as a fuel. I suggested that it might be somewhat toxic and a bit dangerous to synthesize and handle, but he assured me that it was (a) very easy to put together, and (b) as harmless as mother's milk. I was dubious, but told him that I'd see what I could do.

I looked the stuff up, and discovered that, indeed, the synthesis was easy, but that it was extremely toxic, and a long way from harmless. As I had suffered from mercury poisoning on two previous occasions and didn't care to take a chance on doing it again, I thought that it would be an excellent idea to have somebody else make the compound for me. So I phoned Rochester, and asked my contact man at Eastman Kodak if they would make a hundred pounds of dimethyl mercury and ship it to NARTS.

I heard a horrified gasp, and then a tightly controlled voice (I could hear the grinding of teeth beneath the words) informed me that if they were silly enough to synthesize that much dimethyl mercury, they would, in the process fog every square inch of photographic film in Rochester, and that, thank you just the same, Eastman was *not* interested. The receiver came down with a crash, and I sat back to consider the matter. An agonizing reappraisal seemed to be indicated.

Phil wanted density. Well, dimethyl mercury was dense, all right—d = 3.07—but it would be burned with RFNA, and at a reasonable mixture ratio the total propellant density would be about 2.1 or 2.2. (The density of the acid-UDMH system is about 1.2.) That didn't seem too impressive, and I decided to apply the *reducto ad absurdum* method. Why not use the densest known substance which is liquid at room temperature—mercury itself? Just squirt it into the chamber of a motor burning, say, acid-UDMH. It would evaporate into a monatomic gas (with a low C_P, which would help performance), and would go out the nozzle with the combustion products. *That* technique should give Phil all the density he wanted! Charmed by the delightful nuttiness of the idea, I reached for the calculator.

For my calculations I used the monopropellant Cavea A, not only because it had a good density by itself (1.5) but because it would be simpler to handle two liquids than three in the wildly improbable event that things ever got as far as motor work. I calculated the performance of Cavea A with various proportions of mercury—up to six times the mass of the primary propellant. (It was easy to fit mercury into the NQD calculation method.) As expected, the specific impulse dropped outrageously as mercury was added to the system, but the density impulse (specific impulse × propellant density) rose spectacularly, to peak at 50 percent above that of the neat monopropellant with a mercury/propellant ratio of about 4.8.

The next thing was to set up the boost velocity equation: $c_b = c \ln (1 + \phi d)$, and to plug in the results of the performance calculations. I did this for various values of ϕ,[*] plotting the percentage increase in boost velocity over that produced by the neat propellant against the percentage of the (fixed) tank volume filled with mercury rather than propellant. The result was spectacular. With $\phi = 0.1$, and 27.5 percent of the tank volume filled with mercury instead of propellant, the bulk density was 4.9 and the boost velocity was about 31 percent above that of the neat propellant; at $\phi = 0.2$ there was a 20 percent increase with 21 volume percent of mercury. At $\phi = 1.0$, on the other hand, the best you could get was a 2 percent increase in boost velocity with 5 volume percent of mercury. Obviously, a missile with a low ϕ, such as an air-to-air job, was where this system belonged—if anywhere.

I solemnly and formally wrote the whole thing up, complete with graphs, labeled it—dead pan—the "Ultra High Density Propellant Concept," and sent it off to the Bureau. I expected to see it bounce back in a week, with a "Who do you think you're kidding?" letter attached. It didn't.

Phil bought it.

---

[*] ϕ, as you may remember, is a loading factor: the propellant tank volume divided by the dry mass (all propellants gone) of the missile. If there are ten kilograms of dry mass per liter of tank volume, $\phi = 1/10$, or 0.1.

He directed us, forthwith, to verify the calculations experimentally, and NARTS, horrified, was stuck with the job of firing a mercury-spewing motor in the middle of Morris County, New Jersey.

Firing the motor wouldn't be any problem; the problem lay in the fact that all of the mercury vapor in the atmosphere would not be good for the health of the (presumably) innocent inhabitants of the county—nor for our own. So a scrubber had to be built, a long pipe-like affair down which the motor would be fired, and fitted with water sprays, filters, and assorted devices to condense and collect the mercury in the exhaust before it could get out into the atmosphere. We had it built and were about ready to go, when the Navy decided to shut down—"disestablish"—NARTS, and ordered us to ship the whole mercury setup to NOTS. With a sigh of relief, we complied, and handed them the wet baby. Saved by the bell!

At NOTS, Dean Couch and D. G. Nyberg took over the job, and by March 1960 had completed their experiments. They used a 250-pound thrust RFNA-UDMH motor, and injected mercury through a tap in the chamber wall. And the thing *did* work. They used up to 31 volume percent of mercury in their runs, and found that at 20 percent they got a 40 percent increase in density impulse. (I had calculated 43.) As they were firing in the middle of the desert, they didn't bother with the scrubber. And they didn't poison a single rattlesnake. Technically, the system was a complete success. Practically—that was something else again.

A more practical way to get a high-density system (or so people thought) was to use a metallized fuel, one with a light metal suspended in it. As we have seen, this was an old idea, going back at least to 1929. BMW in Germany tried it about 1944, without noticeable success, and Dave Horvitz at Reaction Motors made a long series of tests, in 1947–51, burning a 10 to 20 percent suspension of powdered aluminum in gasoline, with liquid oxygen. Again, his success was not spectacular. It was difficult to get decent combustion efficiency, and a good part of the metal never burned at all, but was exhausted unchanged out the nozzle. Designing an injector which would handle a suspension wasn't easy, particularly as the viscosity of the suspension varied outrageously with temperature. And if the mixture stood around a while, the aluminum had a strong tendency to settle to the bottom of the tank.

So, although Boeing, in 1953, considered using a suspension of magnesium in jet fuel, and burning it with WFNA (the project never got anywhere) interest in such things languished for some years. What revived it, late in the 50's, was a safety problem.

The Navy had always been reluctant to store loaded liquid rockets in the magazines of its beloved airplane carriers. What would happen if one of them sprung a leak, and disgorged a load of highly corrosive oxidizer, or highly inflammable fuel (or even worse, both of them!) onto the magazine deck? The point being, of course, that below decks on a carrier ventilation is difficult, and furthermore, aboard ship there's no place to run. Somebody—nobody now remembers who it was—came up with the idea that if the propellants were gelled—given the consistency of a not particularly stiff gelatine dessert—leakage would be extremely slow, and the situation would be manageable. As for the problem of injecting a gelled propellant, that could be solved by making the gel thixotropic. Whereupon everybody concerned demanded an explanation of *that* word.

A thixotropic gel, or "thixotrope," is a peculiar beast. Left to itself, it sets up to a comparatively stiff jelly, and if it is pushed gently it resists and flows very slowly, as though its viscosity were very high. If, however, it is subjected to a *large* force, as it would be if shaken violently, or forced under high pressure through an injector, its resistance suddenly collapses as though it had decided to relax and enjoy it, its viscosity drops precipitously, and it flows like a civilized liquid. A thixotropic propellant, then, would reduce the leakage hazard, while still being injectable.*

As it turned out, it wasn't particularly difficult to turn most of the common propellants into thixotropes. Five percent of so of finely divided silica would do it to nitric acid, and the hydrazines could usually be gelled the same way or by the addition of a small percentage of certain cellulose derivatives. And the results *could* be fired, although filling the tank beforehand was a frustrating and infuriating job. Combustion efficiency left something to be desired, and the dead weight of the silica naturally reduced the performance; but the system could be made to work—more or less. The real trouble showed up when an attempt was made to gel the halogen oxidizers. Silica, obviously, was impossible, as were the carbohydrate cellulose type agents. At Aeroprojects they tried to gel a mixture of ClF_3 and BrF_5 with a pyrolytic carbon black, and thought that they had solved the problem, particularly when the gelled mixture showed a card-gap value of zero cards. I was dubious about the whole thing though, and warned their Bill Tarpley and Dana McKinney that the system was inherently unstable, and that they were hunting for trouble. Unfortunately, I was proved right almost immediately. Fred Gaskins, was working with some of the material late in 1959, when it detonated. He lost an eye and a hand, and suffered fluorine burns which would have killed most people. Somehow, he survived, but that was the end of the attempt to mix interhalogens and carbon black. Later attempts used completely fluorinated substances, such as

---

* A jellied, or thixotropic, fuel is much less of a fire hazard than the straight liquid if it *is* spilled. It evaporates and burns much more slowly, and doesn't have a tendency to spread the fire all over the surroundings. Considerable work has been done, recently, on applying the principle to jet fuel in commercial airliners, to reduce the fire hazard in case of a crash.

SbF$_5$, for the gelling agent. Unfortunately, an inordinate amount of the agent was required to do the job.

A few years later, gelling appeared to be the answer to another problem, that of propellant sloshing in space vehicles. If, for some reason, the propellant in a partly full tank starts to slosh back and forth, the center of gravity of the rocket will shift in an unpredictable manner, and directional and attitude control can be lost. A gelled propellant, obviously, isn't subject to sloshing, and in 1965 A. J. Beardell of Reaction Motors, then investigating the diborane/OF$_2$ system for deep space work, looked into the problem of gelling OF$_2$. He found that he could do it with finely divided LiF, which, of course, would not react with the oxidizer. However, since several percent of LiF were needed to form the gel, the performance was appreciably degraded. R. H. Globus of Aerojet discovered a much more elegant solution to the problem three years later. He simply bubbled gaseous ClF$_3$ through liquid OF$_2$. The chlorine trifluoride froze instantly to micropscopic crystals which acted as the gelling agent. Five or 6 percent of the additive made a very fine gel, and the effect on performance was microscopic. For some reason or other, ClF$_5$ wouldn't work.

The gelled propellants revived the interest in metallized fuels. Many people thought that, by gelling a fuel, it might be possible to load it up with 50 percent or so of aluminum, or boron, or perhaps even beryllium—if you could ever get your hands on enough of the last—without having the metal settle out. It was soon discovered, too, that if your metal were finely enough divided, with particle sizes of the order of a micrometer, so that the Van der Waals forces became important, it would itself tend to gel the mixture. So there was a great burst of effort and people all over the country started to investigate the rheological properties of various metallized slurries (these have no gelling agent besides the metal), gels, and even emulsions. (These have two liquid phases—like mayonnaise—besides the metal.) Most of the investigators used Ferranti-Shirley viscosimeters, which can measure the viscosity of such substances as a function of the shear rate. (I was always getting the name confused with "Ferrari," which is not unreasonable, since not only the names, but also the prices were similar.)

These investigators discovered that making a stable gel or slurry was not a science, but a black art, accomplished reliably only with the aid of witchcraft, and that getting two batches of gel with the same rheology was a miracle. But they persisted and in the early 60's several mixtures were ready for test firing.

Boron, aluminum, and beryllium were the metals investigated. Reaction Motors came up with a slurry of boron in a hydrocarbon, intended to be used in a ram-rocket, with chlorine trifluoride as the primary oxidizer. The idea was to maximize the propellant density, and since BF$_3$ is a gas, combustion problems were not serious. Most of the work, however, was directed towards aluminized fuels, and Rocketdyne, as early as 1962, had fired an aluminum-hydrazine mixture with N_2O_4. It contained almost 50 percent of aluminum, and they called it "Alumizine." It was designed for an improved Titan II, but, although they have been working on it ever since, it hasn't yet become operational. Reaction Motors fired an aluminized hydrazine-hydrocarbon emulsion with N_2O_4 two years later, but it, too, has failed to make the grade. And although the Naval Ordnance Test Station has fired their "Notsgel" (aluminum in gelled hydrazine mixtures) successfully many times, it hasn't yet found an application.[*] And there have been other aluminized fuels, but none of them are ready for operational use.

In my own opinion, it will be a long time before they are operational, if they ever are. For the problems are horrendous. They come in two sorts, those arising when you try to store the fuels, and those which show up when you try to fire them, and it's hard to say which resist solution more stubbornly.

A shelf life of five years is specified for a prepackaged missile, and a lot of things can happen to a metallized gel in five years, particularly if the storage temperature varies considerably during that time—as it would if the missile were stored out of doors—or if it is subjected to vibration, which it is certain to be if it is shipped from point A to point B. There is always the tendency for the metal to settle out, and this tendency is abetted by wide temperature variations, which drastically and sometimes irreversibly change the rheology of the gel. And vibration, of course, has a tendency to reduce the viscosity, of a thixotropic gel, temporarily, of course, but possibly long enough to permit appreciable sedimentation. Or syneresis—a peculiar vice to which some gels are addicted—may set in. If this happens, the gel starts to shrink and to squeeze the liquid out of its structure, and the end of the process may be a comparatively small volume of a very dense and stiff solid phase surrounded by a volume of clear liquid. None of these things may happen—but on the other hand, they may—and the state of the art has not advanced to the point at which one can be assured that a metallized gel will survive, unchanged, five years of storage in climates ranging from that of Point Barrow to that of the Mojave Desert.

Most of the gels and slurries which have been considered have been based on hydrazine or hydrazine mixtures, which fact is the cause of another—and very peculiar—problem. Missile tanks are usually made of very pure aluminum. But there are always some impurities, and some of these impurities are likely to be transition metals such as iron which catalyze the decomposition of hydrazine. However, if the concentration of the catalytic metals is only some

---

[*] One of their hydrazine mixtures was a three to one mix of monomethyl hydrazine and ethylene di-hydrazine. This has a freezing point of $-61°$, and the viscosity of the EDH improves the stability of the gel. This is one of the few propellant applications that EDH has found to date.

parts per million, very few of the offending atoms will be on the tank surface itself, where they can make trouble, and the decomposition and gas evolution will be negligible. However, if the hydrazine is loaded with very finely divided aluminum, the surface volume ratio of the metal will be increased by many orders of magnitude, as will be the number of catalytic atoms in contact with the hydrazine. Under these circumstances, the decomposition is increased enormously, and even if it is insufficient to change the composition of the fuel appreciably in a reasonable length of time, the accompanying gas evolution can have serious, and disconcerting, results. For the gas cannot escape from the gel, which thereupon swells up exactly like a cheese soufflé. And try to run *that* through an injector!

Assuming, however, that the storage problems have been coped with, somehow, the operational problems remain. The first of these is that of forcing the fuel out of its tank. If a metallized gel is pressurized—that is, high pressure gas is let into the tank to force the fuel out—a sort of tunneling process takes place. The gas simply blows a hole for its own passage down through the gel to the outlet, and leaves most of the fuel untouched and sitting quietly around the sides of the tank, instead of flowing, as it should, through the feed line to the motor. The fuel has to be completely enclosed, as in a flexible bladder (to which the expulsion pressure is applied), or a large fraction of it simply won't leave the tank. Once the fuel leaves the tank, the rate at which it flows through the fuel line and the injector into the motor is strongly dependent on its viscosity, and the viscosity of a metallized gel varies madly with the temperature. Since the viscosity of the oxidizer doesn't vary nearly as much, the result of this is that the mixture ratio if you fire the motor at −40° will be quite unlike that which you will get if you fire it at +25°—and it certainly won't be the one that you want.

Then, once the fuel is in the motor—and I won't go into the problem of designing an injector which will disperse a gel properly—there's the problem of burning the aluminum. Unless the chamber temperature is well above the melting point of aluminum oxide (about 2050°) or, preferably above the considerably higher temperature at which it decomposes, the aluminum particle will simply coat itself with a layer of solid or liquid alumina, and refuse to burn to completion. When burned with N_2O_4 the chamber temperature is just about high enough to burn an aluminized gel properly. It's highly probable that combustion with nitric acid would be marginal, with a chamber temperature not quite high enough to make the metal burn to completion. (With a halogen oxidizer, such as ClF_3, this particular problem doesn't arise, since AlF_3 is a gas at the temperatures we're talking about.) And, naturally, the dense clouds of solid Al_2O_3 resulting from the combustion of an aluminized gel leave a very conspicuous exhaust trail.

There is one final problem which should be mentioned—final because it comes up when the motor is shut down. The heat from the hot motor soaks back into the injector, and the gel in the injector holes sets up to something resembling reinforced concrete, which has to be drilled out before the motor can be fired again. So, restarts are out of the question.

The problems with beryllium-loaded gels are the same as those with aluminized ones, only more so, and with one or two peculiar to themselves. The exhausted BeO, of course, is violently poisonous, producing something resembling a galloping silicosis, but the most serious problem is in the combustion. Beryllium oxide melts at a considerably higher temperature than does aluminum oxide, and doesn't vaporize until the temperature is near 4000°, so that burning it is even more difficult than burning aluminum. Rosenberg, at Aerojet, burned a beryllium-hydrazine slurry ("Beryllizine") with hydrogen peroxide in 1965, and got a C^* efficiency of some 70 percent, which indicated that *none* of the beryllium had burned. At Rocketdyne, they had the same experience with the combination. When Rosenberg used N_2O_4 as his oxidizer, his C^* efficiency was some 85 percent, showing that *some* of the metal had burned. His performance was particularly bad at what should have been the optimum mixture ratio. Various expedients designed to improve combustion, such as vapor-coating the beryllium powder with chromium, didn't improve the situation appreciably.

Aluminum hydride was a compound that aroused a flurry of interest in the early 60's. It had long been known, but not as pure or relatively pure AlH_3, since it had always been prepared solvated with ether, which couldn't be removed without decomposing the hydride. However, Dow Chemical and Metal Hydrides, late in 1959 or early in 1960, devised methods of obtaining it without ether, and Olin Mathieson soon made important contributions to the synthetic methods. Its intended use was as an ingredient in solid propellants, but the liquid people tried to use it in gels. It wasn't sufficiently stable, but reacted with the hydrazines, evolving hydrogen in the process, so the idea was soon abandoned.

Beryllium hydride, BeH_2 had more staying power. It had been known since 1951, but again, in an impure state. In 1962, however, G. E. Coates and I. Glocking of the Ethyl Corporation managed to prepare it in a fairly pure (about 90 percent) state. It, too, was intended for solid propellant use. It was nicknamed "Beane" (pronounced "beany"), as a security measure. (A little later it was discovered that its stability could be improved by heating it, and the result was called "Baked Beane.") But code name or no, the secret was soon out. I was in Dick Holzmann's office in the Pentagon when an assistant came in with the latest issue of *Missiles and Rockets*. And there was BeH_2, spread all over the page. It appears that a congressman who wanted to show how knowledgeable he was had blown security and had told a reporter everything he knew. I have heard—and used—some spectacular language in my time, but Holzmann's remarks were a high point in the history of oral expression.

Naturally, the liquid people had to see if BeH_2 could be used in a gel. It appeared to be much more stable than aluminum hydride, particularly when it was in the amorphous, rather than the crystalline, state. Rocketdyne reported that the former reacted very little even with water. Texaco, Aerojet, and Rocketdyne investigated it in mono-methyl hydrazine gels between 1963 and 1967. Aerojet claimed the mixture was stable, but Rocketdyne's gel, which had some straight hydrazine in it, displayed the soufflé syndrome. Its longtime stability in hydrazine appears very doubtful; certainly it is thermodynamically unstable.

With liquids which do not have active hydrogens, the situation is different. Grelecki at Reaction Motors, in 1966, made a 55 percent slurry of BeH_2 in dodecane, and burned it with hydrogen peroxide, getting good combustion and a high C^* efficiency. That same year the Ethyl Corporation made an apparently stable slurry of the material with pentaborane, and Gunderloy, at Rocketdyne, has investigated mixtures of the hydride with his beryllium semiliquids.

However, even if they are stable—and not all counties have been heard from—BeH_2 gels and slurries don't appear to be the wave of the future. The toxicity of the exhaust and the high price of the propellant appear to rule them out as far as tactical missiles are concerned, and there doesn't seem to be any other role for them that can't be filled better by something else.

A rather far-out concept, even in the fields of gels and slurries and monopropellants, is that of the heterogeneous monopropellant—a solid fuel slurried or gelled in a liquid oxidizer. The Midwest Research Institute came up with the first of these in 1958, when they suspended powdered polyethylene in RFNA. Unfortunately, its sensitivity was more than 120 cards, and it was thermally unstable to boot, so it was hurriedly abandoned before anybody got hurt. About five years later Reaction Motors introduced a similar mixture, with boron carbide, B_4C suspended in a special high density RFNA containing about 40 percent N_2O_4. This was insensitive to the card-gap test, but was thermally unstable, and it, too, had to be junked. In 1965 they tried mixing boron carbide with ClF_5 (!), and found that it was apparently stable at 65°, although there was some reaction at first when the two compounds were mixed. Nevertheless, apparently remembering what had happened to Fred Gaskins, they didn't carry their experiments any further. And for some years Rocket Research Co., a small organization in Seattle, has been industriously plugging "Monex," a mixture of powdered aluminum, hydrazine, hydrazine nitrate, and water, and, ignorant of or ignoring the work on hydrazine and hydrazine nitrate done nearly twenty years before at NOTS, claiming an outstanding and original contribution to rocketry. Recently they have been experimenting with beryllium instead of aluminum. Combustion efficiency with these propellants, particularly the beryllium-based ones, is bound to be bad, since the chamber temperature is comparatively low. Rocketdyne, in 1966, did some work with similar beryllium mixtures, with no notable success. The heterogeneous monopropellants can only be considered an aberration, off the main line of propellant development, and highly unlikely ever to lead to anything useful. About all it proves is the willingness of rocket people to try *anything*, no matter how implausible, if they can con NASA or one of the services into paying for it.

This may explain the work on the "Tribrid" (an etymological monstrosity, if there ever was one!). These are propellant systems involving three propellants, and the name derives vaguely from "hybrid." Sometimes the term "tri-propellant" is used. Performance calculations made in the early 60's showed that for space use, there were two propellant systems whose specific impulses exceeded those of any other system that could be dreamed up—and exceeded them by a spectacular margin. The first of these was the Be-O-H system, in which the beryllium was burned to BeO by the oxygen, and the hydrogen provided the working fluid. It started to arouse considerable interest in 1963 or so, and Atlantic Research and Aerojet started programs designed to prove it out.

Atlantic Research's approach was an extension of the hybrid system. Powdered beryllium was made into a solid grain with the help of a small amount of hydrocarbon binder. This was burned, as in a hybrid, with the oxygen, and then hydrogen was fed into the chamber downstream of the grain. (In a variant arrangement, some of it was introduced upstream with the oxygen, and the rest farther down.) A scrubber was needed, of course, to take the BeO out of the exhaust stream—and the totality of the precautions taken to avoid poisoning the bystanders was fantastic. In any case, although the motor could be and was fired, combustion efficiency was extremely poor, and the system never, practically, approached its theoretical potential.

G. M. Beighley, at Aerojet, tried another approach, this one resembling the usual bi-propellant arrangement. His two propellants were liquid hydrogen and a slurry of powdered beryllium metal in liquid oxygen. He was able to report his results by 1966, and they were not encouraging. He never got more than 70 percent combustion efficiency, and was plagued by "burnbacks" of his Be-O$_2$ slurry through the injector. It's really surprising that he didn't manage to kill himself.

At any rate, he didn't continue the work, and as little has been heard of the Be-H-O system in the last few years, it is probably dead. When the combustion difficulties are added to the toxicity of BeO and the price of beryllium, there isn't really much point in continuing with it.

The Li-F-H system looks much more promising, and has been investigated rather thoroughly by Rocketdyne. Here, two approaches are possible. Lithium has a low melting point for a metal—179°—so it is possible to inject lithium, fluorine, and hydrogen into the motor, all as liquids, in a true tripropellant system. Or, the lithium can be slurried in the hydrogen, so that the motor can be run as a bi-propellant system. Rocketdyne started investigating Li-H₂ gels in 1963, and three years later Bill Tarpley and Dana McKinney of Technidyne (Aeroprojects renamed) reported gelling liquid hydrogen with lithium and with lithium borohydride. Satisfactory and stable gels were produced with 61.1 weight percent (17.4 volume percent) of lithium or 58.8 weight percent (13.3 volume percent) of lithium borohydride. The evaporation rate of the hydrogen was reduced by a factor of 2 or 3, and gelling the fuel eliminated the propellant sloshing problem.

Their work was, however, only on the liter scale, and in the meantime Rocketdyne went ahead with the other approach, and fired the combination in a true tripropellant motor. They used liquid lithium and liquid fluorine, but used gaseous hydrogen instead of liquid. I presume that they considered that handling two such hairy liquids as fluorine and lithium at the same time was enough, without adding to their misery by coping with liquid hydrogen. I have described some of the problems associated with liquid fluorine, and liquid lithium has its own collection of headaches. You have to keep it hot, or it will freeze in the propellant lines. You must also keep it from contact with the atmosphere, or it will burst into brilliant and practically inextinguishable flame. Add to this the fact that liquid lithium is highly corrosive to most metals, and that it is incompatible with anything you might want to use for gaskets and sealing materials (it even attacks Teflon with enthusiasm), and you have problems.

But somehow the Rocketdyne crew (H. A. Arbit, R. A. Dickerson, S. D. Clapp, and C. K. Nagai) managed to overcome them, and made their firings. They worked at 500 psi chamber pressure, with a high expansion nozzle (exit area/throat area = 60) designed for space work. Their main problem stemmed from the high surface tension of liquid lithium, orders of magnitude higher than that of ordinary propellants, which made it difficult to design an injector that would produce droplets of lithium small enough to burn completely before going out the nozzle. Once this problem was overcome, their results were spectacular. Using lithium and fluorine alone (no hydrogen) their maximum specific impulse was 458 seconds. But when they proportioned the lithium and fluorine to burn stoichiometrically to LiF, and injected hydrogen to make up 30 percent of the mass flow, they measured 542 seconds –probably the highest measured specific impulse ever attained by anything except a nuclear motor. And the chamber temperature was only 2200 K! Performance like that is worth fighting for. The beryllium-burning motor is probably a lost cause, but the lithium–fluorine–hydrogen system may well have a bright future.

13

What Happens Next

The absolute limit to the performance of a chemical rocket, even in space, appears to be somewhere below 600 seconds. This is a frustrating situation, and various far-out methods of cracking this barrier have been suggested. One is to use free radicals or unstable species as propellants, and to use the energy of their reversion to the stable state for propulsion. For instance, when two atoms of hydrogen combine to form one molecule of H_2, some 100 kilocalories of energy are released per two-gram mole. This means that a 50–50 (by weight) mixture of monatomic hydrogen and ordinary hydrogen would have a performance of some 1000–1100 seconds. That is, it would if (A) you could make that much monatomic hydrogen and could mix it with ordinary hydrogen and (B) if you could keep it from reverting immediately to H_2—in a catastrophic manner. So far, nobody has the foggiest idea of how to do either one. Free radicals such as CH_3 and OOF *can* be made, and can be trapped in a matrix of, say, frozen argon, whose mass is so great compared to that of the captured radicals that the whole idea is a farce as far as propulsion is concerned. Texaco for one has been investigating such trapping phenomena and the electronic states in the trapped molecular fragment for several years, but the whole program, interesting as it is academically, must be classified as a waste of the taxpayers' money if it is claimed to be oriented toward propulsion. To quote a mordant remark heard at one meeting, "The only people who have had any luck at trapping free radicals are the FBI."

So it appears that the only practical way to increase the specific impulse in large-thrust applications is to shift to the nuclear rocket, which, fortunately, works and is well on the way toward operational status. (Ionic and other electrical thrusters are useful only in low-thrust applications, and are outside the scope of this book—and of my competence to describe them.) So the chemical rocket is likely to be with us for some time.

And here are my guesses as to which liquid propellants are going to be used during the next few years, and possibly for the rest of the century, although here I'm sticking my neck out a long way.

For short-range tactical missiles, with a range up to 500 km or so, it will be RFNA-UDMH, gradually shifting over to something like ClF_5 and a hydrazine-type fuel. Monopropellants are unlikely to be used for main propulsion, and the problems with gels and slurries are so great that it is unlikely that the benefits to be derived from them can outweigh the difficulty of developing them to operational status.

For long-range strategic missiles, the Titan II combination, N_2O_4 and a hydrazine mixture will continue in use. The combination is a howling success, and if somebody wants to put a bigger warhead on the brute—I can't see why—it would be a lot simpler just to build a bigger Titan than to develop a new propellant system.

For the big first-stage space boosters we will continue to use liquid oxygen and RP-1 or the equivalent. They work and they're cheap—and Saturn V uses a lot of propellant! Later, we *may* shift to hydrogen as a first-stage fuel, but it appears unlikely. The development of a reusable booster won't change the picture, but if a ram-rocket booster is developed all bets are off.

For the upper stages, the hydrogen-oxygen combination of the J-2 is very satisfactory, and will probably be used for a long time. Later, as more energy is needed, there may be a shift, for the final stage, to hydrogen-fluorine or hydrogen-lithium-fluorine. The nuclear rocket will take over there.

For lunar landers, service modules, and the like, N_2O_4 and a hydrazine fuel seem likely to remain useful for a long time. I can't think of any combination likely to displace them in the foreseeable future.

Deep space probes, working at low temperatures, will probably use methane, ethane, and diborane for fuels, although propane is a possibility. The oxidizers will be OF_2, and possibly ONF_3 and NO_2F, while perchloryl fluoride, ClO_3F, would be useful as far out as Jupiter.

I see no place for beryllium in propulsion, nor any role for N_2F_4 or NF_3. Perchloryl fluoride may, as I've mentioned, have some use in space, and as an oxygen-bearing additive for ClF_5, which will probably displace ClF_3 entirely. Pentaborane and decaborane and their derivatives will, as far as liquid propulsion is concerned, revert to their former decent obscurity. Hydrogen peroxide will continue to be used, as a monopropellant, for attitude control and in other low-thrust applications. It will probably not be used as an oxidizer for main propulsion.

This is the picture, as I see it in my somewhat clouded crystal ball. It may be wrong in detail, but I believe that, on the whole, it won't appear too far out of drawing twenty years from now. There appears to be little left to do in liquid propellant chemistry, and very few important developments to be anticipated. In short, we propellant chemists have worked ourselves out of a job. The heroic age is over.

But it was great fun while it lasted.

Glossary

Note. Temperatures are given in degrees Celsium (Centigrade) unless otherwise specified.

A "Compound A," ClF_5.

A-4 German ballistic missile that was used to bombard London, also called V-2.

AN Ammonium nitrate, or amine nitrate monopropellant.

ARIB Aeronautical Research Institute, at Braunschweig.

ARPA Advanced Research Projects Administration.

ARS American Rocket Society. Joined with the Institute of Aeronautical Sciences to form the American Institute of Aeronautics and Astronautics, AIAA.

BECCO Buffalo Electrochemical Company.

BMW Bavarian Motor Works.

BuAer Bureau of Aeronautics, U.S. Navy. Later combined with Bureau of Ordnance, BuOrd, to form Bureau of Weapons, BuWeps.

CTF Chlorine tri fluoride, C_1F_3.

EAFB Edwards Air Force Base, in the Mojave Desert, California.

EES Engineering Experiment Station, Annapolis (Navy).

ERDE Explosives Research and Development Establishment, at Waltham Abbey, England.

Flox Mixture of liquid oxygen and liquid fluorine. A number following, as in Flox 30, indicates the percentage of fluorine.

FMC Food Machines and Chemical Company.

GALCIT Guggenheim Aeronautical Laboratory, California Institute of Technology.

GE General Electric Company.

ICBM Intercontinental Ballistic Missile.

IITRI Illinois Institute of Technology Research Institute, formerly the Armour Research Institute.

IR Infra Red.

IRBM Intermediate Range Ballistic Missile.

IRFNA Inhibited Red Fuming Nitric Acid.

IWFNA Inhibited White Fuming Nitric Acid.

JATO Jet assisted take-off—rocket for boosting overloaded airplanes into the air.

JP Jet Propellant, kerosene type. A number following, as JP-4, indicated a particular specification.

JPL Jet Propulsion Laboratory, Pasadena, operated by Cal Tech.

λ (Lambda) A measure of the oxygen balance in a propellant or combination of propellants. $\lambda = (4C + H)/2O$, where C, H, and O are the number of moles of carbon, hydrogen and oxygen in the combination, and equals the ratio of reducing to oxidizing valences.

LFPL Lewis Flight Propulsion Laboratory, Cleveland, a NACA-NASA facility.

LOX Liquid oxygen.

LRPL See NARTS.

MAF Mixed amine fu el. Number following indicates type. Reaction Motors mixture.

MHF Mixed hydrazine fuel. Number following indicates type. Reaction Motors mixture.

MIT Massachusetts Institute of Technology.

MMH Monomethyl hydrazine.

MON Mixed oxides of nitrogen, $N_2O_4 + NO$. Number following indicates percentage NO.

NAA North American Aviation.

NACA National Advisory Council on Aeronautics. Became NASA.

NARTS Naval Air Rocket Test Station, Lake Denmark, Dover, New Jersey. In 1960 taken over by Army, and became Liquid Rocket Propulsion Laboratory, LRPL, of Picatinny Arsenal.

NASA National Aeronautical and Space Administration.

NOL Naval Ordnance Laboratory, Silver Spring, Maryland.

NOTS Naval Ordnance Test Station, China Lake, California, often referred to as Inyokern.

NPN Normal Propyl Nitrate.

NUOS Naval Underwater Ordnance Station, formerly Naval Torpedo Station, Newport, Rhode Island.

NYU New York University.

O/F The ratio of the oxidizer flow to the fuel flow in a liquid rocket.

ONR Office of Naval Research.

PF Perchloryl Fluoride, ClO_3F.

R&D Research & Development.

RFNA Red Fuming Nitric Acid.

RMD See RMI.

ROR Rocket on Rotor, used to improve the performance of a helicopter.

RMI Reaction Motors, Inc., later became RMD, Reaction Motors Division of Thiocal Chemical Co. Died late in 1969.

SAM Surface to Air Missile.

SFNA Stable Fuming Nitric Acid (obsolete).

Tonka German rocket fuels based on xylidines.

TRW Thompson Ramo-Wooldridge Corporation.

UDMH Unsymmetrical dimethyl hydrazine.

UFA German moving picture company of the 1920's and 1930's.

USP United States Pharmacopea.

UTC United Technology Corporation, a subsidiary of United Airlines.

V-2 The propaganda name for A-4.

VfR Verein fur Raumschiffart, the old German Rocket Society.

Visol. German rocket fuels based on vinyl ethers.

WADC Wright Air Development Center, Dayton, Ohio; Air Force installation.

WFNA White Fuming Nitric Acid.

Index

A

A. *See* Compound A

A-4. *See* German ballistic missile (A-4)

Abramson, Bert, 67, 127, 130–131

Acetic anhydride, WFNA freezing point and, 44

Acetonitrile
  decaborane plus hydrazine and, as monopropellant, 146–147
  as hydrazine additive, 39, 146–47
  tetranitromethane plus, as monopropellant, 135

Acetylene, 30
  liquid hydrogen plus, as monopropellant, 134n
  oxygen-ammonia system and, 95
  research with derivatives, 29, 30, 30–31, 122

Acid(s)
  ignition and reactions with bonds, 26
  See also specific acid

Advanced Research Projects Administration (ARPA), "Project Principia," 153, 158, 160

Aerojet Engineering
  founding, 18
  research, 19, 35, 37, 49, 56, 97; borane, 113, 117; borohydrides, 117; deep space rockets, 78; hybrid propellants, 160, 166, 169, 170, 171; hydrazines, 117; liquid fluorine, 100; liquid hydrogen, 97; monopropellants, 125, 132, 139, 147, 158; oxidizers, 74

Aeronautical Research Institute, Braunschweig (ARIB), 10

Aeroprojects, hybrid propellant research, 165, 172

Aerozine-50, 39

Agena motor, fuel for, 56

Ahlert, Bob, 146, 150

Air Products, monopropellant research, 134

Air Reduction Company
  monopropellant research, 122, 149–150
  ozone work, 103–104
  propellant production, 32

Alcohol(s)
  additives and heat flux, 97
  aniline and RFNA plus, use of, 22
  hypergolicity studies, 25
  liquid oxygen plus, use of, 94, 97, 107
  for long-range ballistic missiles, 21
  See also specific alcohol

Alkyl borane derivatives, research on, 112, 113

Allied Chemical and Dye Company
  research, 45
  monopropellants, 133, 154
  oxidizers, 76, 77

Altman, Dave, 19

Aluminum additives, 7
  in gasoline, 14
  in gelled propellants, 166–167; problems with, 167–168, 169
  for high-density fuels, 164

About the Author

JOHN D. CLARK joined the Naval Air Rocket Test Station at Dover, N.J., in 1949. (It became the Liquid Rocket Propulsion Laboratory of Picatinny Arsenal in 1960.) He retired in 1970, after having been in charge of liquid propellant development for most of the twenty-one years.

A native of Alaska, Dr. Clark attended the University of Alaska and received the B.S. from the California Institute of Technology, the M.S. from the University of Wisconsin, and, in 1934, the Ph.D. from Stanford University. He worked as a chemist in several industries prior to his association with NARTS.

A great deal of Dr. Clark's writing has been in the form of government reports which are classified, but he has also contributed scientific articles and papers in the open literature and, as an avocation, written stories and articles for science fiction magazines.